Debate about the Earth

DEBATE ABOUT THE EARTH

Approach to Geophysics through Analysis of Continental Drift

H. TAKEUCHI, *Professor of Geophysics, University of Tokyo*

S. UYEDA, *Associate Professor of Geophysics, University of Tokyo*

H. KANAMORI, *Associate Professor of Geophysics, University of Tokyo*

Translated by Keiko Kanamori
Illustrated by James K. Levorsen

FREEMAN, COOPER & CO.
Stockton Street, San Francisco 94133

Based on Science of the Earth by
Takeuchi and Uyeda, Nippon Hoso
Publishing Co., 1964

Printed in the United States of America
Library of Congress Catalogue Card Number 67-21261

Acknowledgments

Acknowledgment is gratefully made to the following authors and publishers for permission to quote from copyrighted material: Academic Press Inc., paragraph 3, p. 315, and paragraph 1, p. 316, of *Continental Drift*, S. K. Runcorn (ed.), 1962. American Association of Petroleum Geologists, paragraph 1, p. 76, paragraph 1, p. 111, paragraph 2, p. 194, of *Theory of Continental Drift, A Symposium*, 1928. Cambridge University Press, last paragraph, p. 367, paragraph 1, p. 368, of *The Earth*, 4th ed., H. Jeffreys, 1959. E. P. Dutton & Co., New York, and Methuen & Co., London, paragraphs 2 and 3, p. 5, and paragraph 1, p. 8, of *The Origin of Continents and Oceans*, A. Wegener, 1924. Ronald Press, New York, and Thomas Nelson & Sons, London, last paragraph, p. 508, of *Principles of Physical Geology*, A. Holmes, 1945. S. K. Runcorn, paragraph 8, p. 289, of "Rock Magnetism— Geophysical Aspects," *Phil. Mag. Supplement*, vol. 4, 1955. The Royal Society, London, paragraphs 3 and 5, p. ix, of *A Symposium on Continental Drift*, 1965. J. T. Wilson, paragraph 3, p. 16, paragraphs 1 and 2, p. 17, of "The Movement of Continents," *Symposium on the Upper Mantle Project*, International Union of Geodesy and Geophysics, XIII General Assembly, Berkeley 1963.

Preface

In 1912, Alfred Wegener, German scientist, advanced the theory that some of the continents must have drifted over the surface of the earth to their present positions. For example, two hundred million years ago, according to Wegener, the Atlantic Ocean did not exist, the two American continents being then joined to Europe and Africa.

The idea that drifting of such huge land masses over thousands of kilometers could have happened seemed fantastic. Heated controversy over the theory arose among geologists and geophysicists. In addition, the very boldness of the idea attracted the interest of the general public. The controversy raged for some twenty years.

Then Wegener, in 1930, met a tragic death in the course of an expedition in Greenland. The combination of his death and of the inability to prove or disprove his theory led to a stalemate and then to indifference. Debate about the theory was put aside and apparently Wegener's idea was no longer a part of geologic thinking.

Problems of geology involve such large masses of material and such complex data that a geologist must work persistently over a long period to acquire the pertinent information, to coordinate it, and only then to seek to interpret. Many of the basic theories of geology formulated earlier in this century and now widely accepted were the results of long years of preliminary work going back to the late 18th century. Also, geophysics—the result of the merger of geology and physics—was a young science when Wegener first spoke.

It is not surprising, given the foregoing state of affairs, that Wegener's theory, so revolutionary in concept, struck geologic thinking like a bombshell, and that it could not be quickly proven or disproven. It follows that it is not surprising that, with no decisive conclusion and Wegener's death, the theory was set aside and apparently forgotten. However, in the 1950's, Wegener's idea again exploded in the face of science.

Geophysicists in England, such as S. K. Runcorn and P. M. S. Blackett, studying the earth's magnetism, rekindled the old controversy. They suggested that the solutions to some of the questions their studies raised might be found in the idea of continental drift. By now, geophysics had made great strides in developing techniques and collecting data.

At present, we authors are not necessarily convinced that the theory of continental drift is correct. We do say that various discoveries made in the past few years stand favorable to this theory. The purpose of this book, then, is not to present merely the case for or against the theory. Rather, we propose to study this very engrossing idea in the light of modern knowledge and techniques of geophysics and geology. Also this book will show how various fields of earth science, such as paleontology, rock magnetism and oceanography, which are apparently unrelated to each other, do in fact have important interconnections when we come to study a fundamental problem such as continental drift.

The contents of the book are as follows.

Chapter 1 outlines Wegener's continental drift theory, the evidences for it, and some of the earth's riddles that, according to Wegener, can be successfully explained by the theory. A number of basic facts of geophysics are introduced at pertinent points.

Chapter 2 discusses the controversy over the theory—the major points of the objections and the rebuttals. The debate being inconclusive, the theory is set aside and gradually forgotten until it is revitalized in the 1950's by fresh evidences from the study of the earth's magnetism.

Chapter 3 explains some fundamental facts of the earth's magnetism and discusses various hypotheses on its origin.

Chapter 4 deals with the "fossils" of the geomagnetic field preserved in rocks. Possible reversals of the earth's field in the past are discussed.

Chapter 5 describes how the study of rock magnetism led to the revival of the continental drift theory.

Chapter 6 is devoted to the thermal history of the earth and to the significance of the drift theory in terms of it.

Chapter 7 introduces some recent discoveries in the study of the ocean floor. These discoveries favor the existence of thermal convection in the mantle. Of all the mechanisms for continental drift that have been proposed so far, thermal convection in the mantle is the one which today receives the most favorable attention and critical study.

Chapter 8 contains brief concluding remarks.

The bibliography is mainly limited to easily available books and review articles. Through these, the interested reader can refer to the original papers.

The Japanese edition of this book grew out of a television program "Earth Science," given by one of the authors in an educational program of the Japan Broadcasting Company (NHK). It was published as one of the NHK books, by Nippon Hoso Publishing Company.

The authors wish to express their thanks to Dr. Chuji Tsuboi, professor emeritus of the University of Tokyo, and Dr. Akiho Miyashiro of the Geological Institute of the University of Tokyo for reading the manuscript of the Japanese edition and making helpful suggestions.

We owe Dr. George Thompson of the Department of Geophysics of Stanford University thanks for introducing us to Mr. William H. Freeman.

The book was translated by Mrs. Keiko Kanamori, with the aid of our good friend the geophysicist, Dr. Hiroo Kanamori. As the contents of the book were considerably expanded by Dr. Kanamori, we wished him to be a co-author in the English edition. The translation and the expansion seem to us most satisfactory.

Special thanks are due to Mr. William H. Freeman who showed an unfailing enthusiasm throughout the preparation of the English edition, and made some valuable suggestions for its expansion.

October 1966   H. TAKEUCHI
                                                 S. UYEDA

Translator's Note

I am grateful to Dr. Allan Cox of the U.S. Geological Survey for reading a part of the manuscript and making valuable suggestions. The translation was much improved by stylistic revisions by Mr. William H. Freeman and Mr. John F. Gallagher of Freeman, Cooper & Company to whom I wish to express my gratitude. Thanks are also due to Mrs. Nancy Anderson and Miss Margaret Brown for their help in smoothing away some language difficulties. Finally, I am much indebted to my husband whose constant encouragement and help made the translation of the book possible.

October 1966   K. KANAMORI

Contents

Rifts of the earth—Are continents being rifted apart?—Is the Atlantic a rift?—Speed of the belt conveyer—The current in the Pacific—Below the Japanese Islands—A remark on the mantle convection theory

List of Illustrations

List of Tables

1

Drifting Continents

Once, There Was No Atlantic

Look at the east coast of South America and the western coastline
of Africa in Figure 1-1. Mark the striking resemblance between the

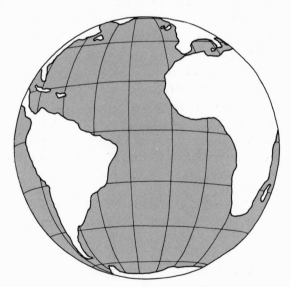

FIGURE 1-1.

two. Alfred Wegener, a young scientist teaching meteorology at
Marburg, Germany, hit upon an explanation for this in 1910. It

occurred to him that South America and Africa might once have formed a single continent that might have split, the parts drifting apart. Few ideas have been bolder and more fantastic.

Moreover, Wegener believed that, by assuming that continents have drifted, many geological phenomena could be explained more satisfactorily than they ever had been. Wegener then turned from meteorology to geology and paleontology and through his study of these worked out what is known today as the "Theory of Continental Drift." He set it forth in his monumental book, *Die Entstehung der Kontinente und Ozeane*, published at Brunswick, Germany, in 1915. An English translation from the third German edition, *The Origin of Continents and Oceans*, was published in 1924 (*21*).

The Man Wegener

Wegener, born in Berlin in 1880, started his career as a meteorologist. His interest in the thermodynamics of the atmosphere—the conditions in upper and polar air masses that "make" the weather —led him in 1906 to join the Danish expedition to northeast Greenland where he spent two winters in research. At the Physical Institute at Marburg, Germany, where he taught from 1908 to 1912, he is remembered as a vigorous young tutor whose outstanding characteristics were intellectual frankness and open-mindedness, coupled with a modesty in his approach to students. His lectures always captivated his audience by their remarkable simplicity and vividness. In 1910, to quote Wegener himself:

> The first notion of the displacement of continents came to me . . .
> when, on studying the map of the world, I was impressed by the
> congruency of both sides of the Atlantic coasts, but I disregarded
> it at the time because I did not consider it probable. In the
> autumn of 1911, I became acquainted (through a collection of
> references which came into my hands by accident) with the
> paleontological evidence of the former land connection between
> Brazil and Africa, of which I had not previously known. This
> induced me to undertake a hasty analysis of the results of research

FIGURE 1-2.

Alfred Wegener. (With permission of United Press International Inc.)

in this direction in the spheres of geology and paleontology, whereby such important confirmations were yielded that I was convinced of the fundamental correctness of my idea . . . Afterwards, the participation in the traverse of Greenland under J. P. Koch of 1912/13 and, later, war-service hindered me from further elaboration of the theory. In 1915, however, I was able to use a long sick-leave to give a somewhat detailed description of the theory. . . . (*21*)

In 1919, Wegener moved to the Meteorological Experimental Station of the German Marine Observatory at Grossborstel, north of Hamburg. Here, in his primitive workroom, he could devote himself to the further elaboration of his theory. J. Georgi, his close associate of those days, recalls:

> The working out of this (continental drift theory) was now progressing well. After the end of the war, with the renewal of contacts with the rest of the world, there came not only news from colleagues . . . but also these experts from all over the world came in person to visit the modest wooden huts in Grossborstel or the nearby Köppen . . . Wegener's house. At that time, one could regard Grossborstel as the Mecca of geophysicists and ecologists interested in this problem. . . . (*16*)

Wegener's fame as the initiator of the continental drift theory should not obscure his contribution to, and lifelong interest in the exploration of Greenland. In 1930, he set off on his fourth expedition. In November of the same year, on his fiftieth birthday, he left the northern-most base in central Greenland for the west coast and never returned.

Geologic Time

Before going into Wegener's theory, we need to know how the chronology of the earth is determined. The history of the earth is immensely long, beyond all comparison with history in its ordinary sense. The history of civilized man goes back no more than several thousand years; the history of the earth goes back billions of years.

Geologists study the history of the earth as told in its layers of rock or "strata." The first law in determining the chronology of the strata is that, if stratum B overlies stratum A, A is older than B. This is self-evident: without the prior existence of A, B would not overlie A. This fact, first recognized by N. Steno (1631–1687) about 300 years ago, is called the *law of superposition*. Though this law makes clear the sequence of stratified rocks in one place, it gives no clue as to the chronological relations between strata at different places.

For such information and correlation, geologists study the fossils of animals and plants preserved in the strata. If fossils of a species of plant or animal or of an assemblage of such species are discovered in a stratum, it follows that the stratum was formed at a period when that particular species or assemblage existed on earth. Therefore, if fossils of the same species are found at two different places, and the species was one that existed for a relatively short time, then the strata containing them must be of approximately the same age. This second law, called the *law of faunal assemblages*, was established by an Englishman, William Smith (1769–1839).

Geological age determined by the study of fossils, which is called paleontology, is divided into *eras* such as Cenozoic ("recent life") and Paleozoic ("ancient life"), which in turn are subdivided into *periods* such as the Carboniferous ("coal bearing") or the Cretaceous ("chalk-like"). Table 1-1 shows the geological time scale in a simplified form.

Source of vast and valuable information though it is, paleontology has its shortcomings. For one thing, it can tell us nothing about strata that contain no fossils. Just as in human history, during which there was a time when, though mankind existed, he was creating no written record, so in earth history there was a time— the Precambrian era—which has left us few tell-tale fossils to reveal much about strata.

Moreover, the paleontological method cannot determine "absolute" dates of geological eras or periods, but indicates only relative ages. For instance, we can guess from the study of evolution that the Jurassic period, the golden age of reptiles, must have occurred very long ago, but whether fifty million or five hundred million years, we cannot tell from the fossil record.

The recent discovery of "radioactive clocks" in rocks has enabled us to overcome this defect of paleontology. Radioactive elements, such as uranium and potassium, disintegrate spontaneously in the rocks and at a steady rate. By making use of this phenomenon, one can determine the "absolute" age of the rocks (see page 211). The figures in the left column of Table 1-1 represent the dates determined by the radioactive method. We now

m.y.* ago	Eras, names† and extent* in m.y. (= million years)	Periods (see also facing page)
Today...	**Cenozoic.** Era of "recent life," extending back 70 m.y.	*Quaternary*..... *Tertiary*.......
50......		
100......	**Mesozoic.** Era of "middle life," extending from 70 to 225 m.y.	*Cretaceous*......
150......		*Jurassic*.......
200......		*Triassic*.......
250......	**Paleozoic.** Era of "ancient life," extending from 225 to 600 m.y.	*Permian*.......
300......		*Carboniferous*...
350......		*Devonian*......
400......		*Silurian*.......
450......		*Ordovician*......
500......		*Cambrian*......
550......		
600......		
↑	**Precambrian.** Era before oldest Paleozoic, extending from 600 to 4500 m.y. This era comprises more than 85% of earth history. Its rocks have yielded relatively little evidence of organic life.	.... No widely accepted system of names or periods exists for Precambrian.
4500.....	*Beginning of the world.*	

Periods (cont.)		Mountain building ages
Extent (m.y.)	Origins of names†	
..... 1	Quaternary ("4th part").	*Alpine orogeny* from present back into late Triassic.
..... 70	Tertiary ("3rd") from 18th Century work.	
..... 65	Name refers to characteristic chalk deposits.	
..... 45	From Jura Mts., Europe, where Period was identified.	
..... 45	"Three-fold" division of period as worked out in Germany.	
..... 45	From Perm, Russia, where Period was identified.	*Variscan orog.* from late Permian back into late Devonian.
..... 80	"Coal-bearing" strata are characteristic.	
..... 50	Period was identified in Devon, England.	
..... 40	Named for Silures, ancient British tribe.	
..... 60	Ordovices, another early British tribe.	*Caledonian orog.* from mid-Silur. back to mid-Cambrian.
..... 100	Cambria was the Roman name for Wales, where this Period's strata were identified.	
.... 3900	re Precambrian, see facing page.	

---

* re Dates and duration of time (m.y. = million years), the figures given are based on data by Arthur Holmes (1959). They are obviously approximations so that one wisely allows 5% more or less. Then the figures fit known facts.

† As to names, they evolved from several hundred years of work, during which systems of naming changed as knowledge grew.

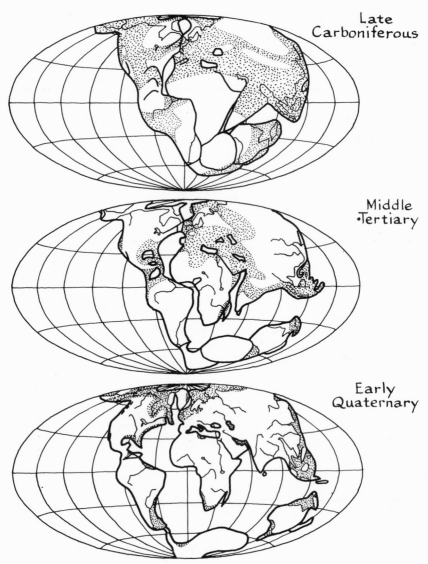

Late
Carboniferous

Middle
·Tertiary

Early
Quaternary

FIGURE 1–3.

Reconstructions of the map of the world for three periods according to
Wegener's theory of continental drift. Dotted areas represent shallow seas.
(After figure 1 in *The Origin of Continents and Oceans*, by Alfred Wegener
(1924), with permission of E. P. Dutton & Co., New York and Methuen
& Co., London.)

therefore know that the span of time since the dawn of the Paleozoic era is about 600 million years, and we can fix the age of the earth at about 4500 million years. Note that the Precambrian era occupies more than 80% of the earth's history. Thus, until the development of the radioactive method, a geologist lacked the tools with which he could study the greater part of the history of the earth.

History of Continental Drift

Figure 1-3 illustrates Wegener's theory of continental drift (21). In these three maps, he intended to show what the world's surface looked like at the end of the Carboniferous period (about 300 million years ago); at the middle of the Tertiary period (about 50 million years ago); and at the beginning of the Quaternary period (about one million years ago), respectively. At the end of the Carboniferous period, North America was joined to the Eurasian Continent and South America to the African Continent. The continents of the Southern Hemisphere, Australia and Antarctica, were also attached to this great mass. The Indian peninsula, longer than it is now, fitted nicely between Africa and Australia. Wegener proposed the name *Pangea* for this great hypothetical continent.

The theory further stated that, during the Jurassic and Tertiary periods, Pangea began to split apart. The continents moved either westward or towards the Equator or both. During the Cretaceous period, South America and Africa, like some gigantic iceberg that had split, began to drift apart, making room for the Atlantic Ocean. Greenland and Norway did not part company until the beginning of the Quaternary period.

Collision of India and the Asian Continent

The Indian peninsula (Figure 1-3) protruded from the Asian Continent in the shape of a tongue, and the root of the tongue, so to speak, was covered by a shallow sea. When Pangea split, India drifted northeastward, compressing the root of the tongue into fold mountains until they formed "the roof of the world," including the lofty snow-clad peaks of the Himalayas. With a little exaggeration, we might say that this means the Himalayas were born out of the collision of India and Asia.

According to Wegener, the front margin of a moving continent eventually encounters the resistance of the ocean floor and is compressed and folded so as to push up mountain ranges. The two Americas in drifting westward formed on their west coasts the magnificent Cordilleras-Andes range, which extends from Alaska to the Antarctica (see Figure 1-4); the mountain ranges of New

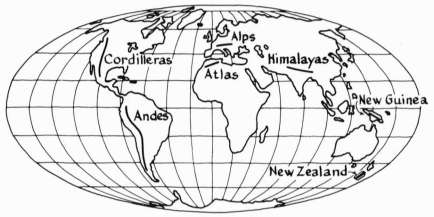

FIGURE 1-4.

Zealand and New Guinea were formed at the front edge of moving Australia. Last, Tertiary fold mountains, such as the Alps and the Atlas in North Africa, can be interpreted as an extension of the Himalayas; they were caught between the continents moving towards the Equator and pressed into fold mountains.

Islands Left Behind

What happens at the rear of a drifting continent? According to Wegener, moving continents may leave in their wakes detached fragments as long strings of islands. For instance, Central America left in its wake the Lesser and Greater Antilles. The tapering ends

of Greenland and South America curve eastward as the result of
the same tendency to lag behind. The island arcs on the east side
of Asia, including Japan and the Philippines, are stragglers which
could not keep pace with the westward movement of that continent.

New Zealand, mentioned in the previous section, is another
interesting example. At first, Australia moved towards New Zealand,
forming fold mountains there; then changing direction, it moved
away from New Zealand, leaving it behind as an island.

Analogy of the Shrivelling Apple

Before Wegener developed his idea of continental drift, the leading
theory on mountain formation had been that of contraction of the
earth. It was assumed that the earth was originally hot and had
gradually cooled. Most substances contract as they cool. With the
cooling and the consequent contraction of the inner part, the surface of the earth wrinkled, as the skin of an apple shrivels when it
dries and contracts. The folding of strata in the Alps or in the
Himalayas, however, was found to be far more intense than could
be expected from surface wrinkling. A schematic drawing of a
folded structure is shown in Figure 1-5.

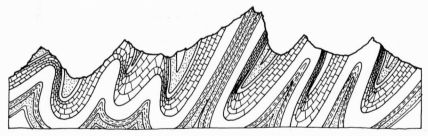

FIGURE 1-5.

An eroded fold.

How is such a folded structure formed? The most plausible
explanation is as follows. Suppose there is a horizontal assemblage

of many strata (Figure 1-6A). The strata become crumpled or folded as the result of horizontal compression (Figure 1-6B). When compression becomes extreme, some lower strata may be pushed over upper strata (Figure 1-6C). Generally, weathering wears and erosion carries away some material; what remains is a mountain range, as we see it today (Figure 1-6D).

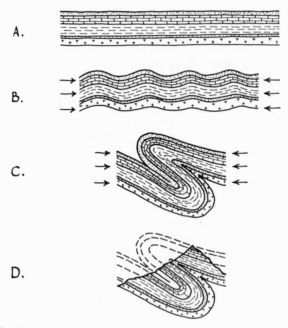

Figure 1–6.

Process of folding.

The intense folding found in the Alps and the Himalayas is called overturned folding, and the horizontal extent of strata has been shortened to one quarter to one eighth of its original length. At that rate, the Alps, at present stretching some 150 kilometers, must once have extended 600 to 1200 kilometers. That contraction in the Alps alone would have meant a shrinking of the circumference of the earth by 3%. To achieve that, as physics tells us, the earth would have had to cool approximately 2400°C. But the earth has other great ranges such as the Himalayas and the Andes. To produce these as well, the earth must have cooled much more than 2400°C. But could the earth once have been so hot and since have cooled off so much? In the 1920's it became necessary to reconsider this question, because it was realized that the radioactive elements contained in the earth, such as uranium, generate a significant amount of heat. Therefore, it seems that the earth as a whole is not cooling: it may even be heating up.

How then were mountains actually formed, if not by cooling of the earth and surface contraction? Wegener's theory offered a simple and logical explanation, a possible solution to this question that the geologists had never answered to their own satisfaction. To comprehend the possibilities of Wegener's theory, we need to know more about the interior of the earth.

Floating Continents

Is a mountain merely a tract of land that stands higher than nearby areas and rests as an excess load on a rigid substratum? No. The structure under a mountain differs from that under other areas. This fact was established during the 19th century through observation of the anomaly of the direction of a plumb line in the Himalayas.

The direction of a string suspending a plumb bob or weight is called the direction of *plumb line*. If a mountain were really a load heaped on a perfectly rigid substratum, the plumb line should deflect toward the mountain because of the attraction of the mountain mass. This is illustrated in Figure 1-7A. However, the actual deflection of the plumb line turned out to be much smaller, about one-third of the expected value (see Figure 1-7B). It follows then that the excess mass of the mountain on the surface must be somehow balanced or compensated for by the structure below. The most plausible way of compensation is this. Look at Figure 1-7B. The mountain mass is made up of lighter material than its substratum and there is a root of the mountain made up of the same lighter material, replacing the heavier substratum. In that case, the attraction of the mountain mass and hence its effect on the deflection of plumb line would be diminished to the observed value.

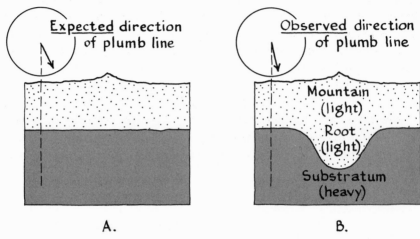

A.   B.

FIGURE 1-7.

Deflection of the plumb line towards a mountain mass. The deflection in this figure is greatly exaggerated. The actual deflection, in the case of the Himalayas, is 10 ~ 30 seconds of arc.

If such compensation generally prevails, the higher the mountain, the deeper its root must penetrate into the substratum. This would be possible if the substratum behaves like a fluid and the lighter mountain mass is floating on the heavier "fluid" substratum, somewhat as copper blocks float in a pan of mercury. Turn to Figure 1-8A. The three copper blocks have the same density but differ in weight because they are of different heights. The tallest block projects highest above the surface and also sinks deepest into the mercury.

This phenomenon can be explained by Archimedes' principle of buoyancy. The principle states that the uplifting effect of a fluid—buoyancy—is equal to the weight of the part of the fluid displaced by the floating body. Each block, placed in mercury,

sinks until buoyancy—the weight of the mercury it displaces—
becomes equal to the weight of the block itself, and is balanced in
that state. Therefore, the tallest block, being the heaviest, must sink
deepest into the mercury. Land masses are considered to float on
their substratum in accordance with the same principle (see
Figure 1-8B). The geologic term for this phenomenon of continental
flotation is *isostasy*.

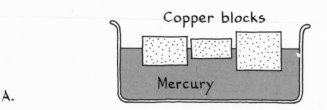

A.

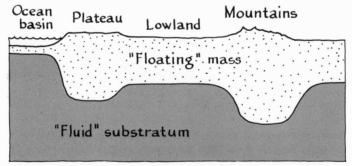

B.

FIGURE 1–8.

The principle of isostasy. (A) Flotation of copper blocks in mercury.
(B) Flotation of land masses in "fluid" substratum.

The idea of isostasy has subsequently been confirmed and
refined by various geophysical evidences, notably seismic data.
Before going into the seismic evidence for isostasy, we will outline
the current view on the structure of the earth's interior.

The Earth's Interior

The earth is often compared to an egg in the arrangement of its layers. The comparison is apt, for the earth is composed of the crust, the mantle, and the core (see Figure 1-9), corresponding to the shell, the white, and the yolk of an egg. The crust and the

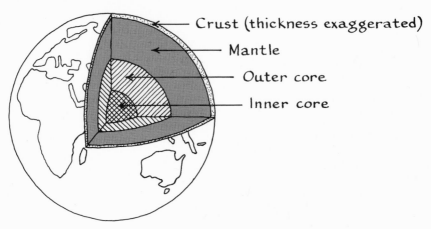

Crust (thickness exaggerated)
Mantle
Outer core
Inner core

FIGURE 1-9.

Cross-section of the earth.

mantle are solid while the outer part of the core is liquid. The deeper part of the core (the inner core) is generally believed to be solid.

Depths of the earth are quite inaccessible to direct human reach; they are more remote, in that sense, than the moon. How do geophysicists find out the state and the composition of the earth at such depths? The best source of information so far is definitely the study of earthquake waves. By studying the propagation of earthquake waves recorded on seismographs, geoscientists can, to some extent, infer the state and the composition of the earth's interior.

Suppose that an earthquake occurred in the vicinity of Tokyo at a shallow depth. The earthquake waves would propagate in all directions, some deep down into and through the earth's interior. These waves from Tokyo would be detected one after another by seismographs set up all over the world, at Hawaii about 9 minutes 40 seconds after the occurrence of the earthquake, at San Francisco about 11 minutes 40 seconds, and at New York about 13 minutes 30 seconds and so on (see Figure 1-10). By working out such data

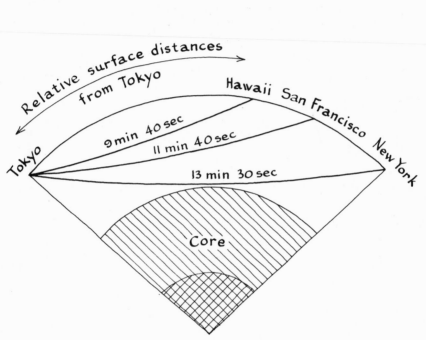

FIGURE 1–10.

Travel time of the P wave.

mathematically, we can determine the speed of the earthquake waves at various depths within the earth.

There are several types of earthquake waves. In the case of a P wave (primary wave), the state of compression and expansion in the matter is transmitted, just as in sound waves (see Figure 1-11A). Another important wave is the S wave (secondary wave).

In this case, the shearing distortion of the material is transmitted (Figure 1-11B). As the resistance of the material to the shearing

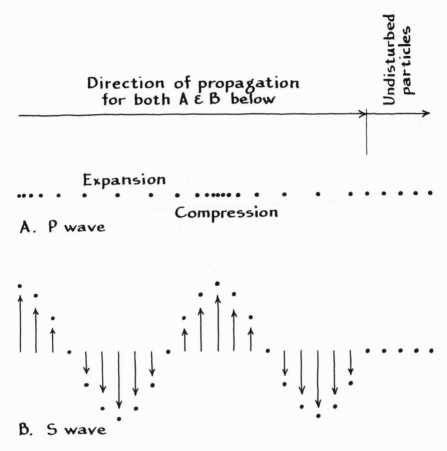

FIGURE 1-11.

The effect of the P and S waves on the grouping of a series of equally spaced particles.

distortion by the wave is an essential factor of transmission, material which shows no resistance to shear, such as liquid, cannot transmit S waves.

In 1909 a Yugoslav seismologist, Mohorovičič, found while studying a Balkan earthquake, that the speed of the P wave increased suddenly some tens of kilometers below the earth's surface. This change in the speed of earthquake waves indicates a significant change of constituent material at this depth. The boundary that marks this abrupt change was named the Mohorovičič discontinuity after its discoverer (or simply the Moho discontinuity). The layer above the Moho discontinuity is called the *crust* and the layer below it the *mantle*.

Recent investigation has revealed that most of the continental crust consists of two layers, the boundary of which is far less distinct than the Moho discontinuity. In the upper layer, the speed of P wave is about 6 km/sec, and the density of the material is about 2.6 ∼ 2.7 g/cm³. Since these properties are close to those of granite, this layer is called the granitic layer. In the lower layer, the speed of P wave is 6.5 ∼ 7 km/sec and the density of the material 2.8 ∼ 3.0 g/cm³. As these properties are similar to those of basalt, the lower layer is called the basaltic layer. The total thickness of the crust, as we shall soon see, varies from place to place, but the average thickness in continental regions is about 35 km. Under oceans, the crust consists only of the basaltic layer, having a thickness of about 5 km.

It is not yet certain what rock composes the mantle right below the Moho discontinuity. Good candidates for the chief constituent rock are peridotite and eclogite. Peridotite is a heavy greenish rock composed mainly of the mineral olivine whose chemical formula is (Mg_2SiO_4); eclogite is a heavy rock containing a large amount of the mineral garnet. Both have higher densities than basalt.

Within the mantle, the speed of earthquake waves increases with depth. Look at Figure 1-12. The speed of the P wave, which is about 8 km/sec just below the Moho discontinuity, increases gradually with depth till it reaches the maximum of about 14 km/sec at the depth of 2900 kilometers. There, however, the speed suddenly drops to 8 km/sec. With the S wave too, the speed increases from about 4.5 km/sec right below the Moho discontinuity to about 7 km/sec at the depth of 2900 kilometers. Beyond this depth, however, no S wave has been known to pass. Since this boundary marks the disappearance of the S wave as well as the abrupt drop in the P wave speed, it was inferred that beyond this depth, the earth must be in a liquid state. This liquid part was

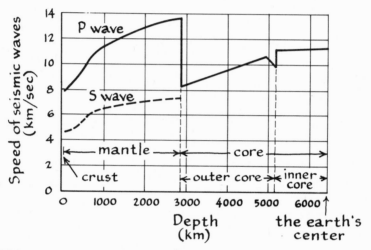

FIGURE 1–12.

Speed of the P and S waves in the earth's interior.

named the *outer core*. It is generally accepted that the outer core consists mainly of liquid iron.

Within the outer core, the speed of the P wave again increases gradually with depth (Figure 1-12). There is an indication that at the depth of about 5000 kilometers, the speed increases abruptly by about 10%. For this reason, the part beyond this depth, called the *inner core*, is usually believed to be solid, though this is still open to question.

A number of seismic evidences have shown that beneath mountains such as the Sierra Nevada and the Alps, the crust is almost twice as thick as the average figure for continental regions; on the other hand, the crust under depressed regions like the ocean basin is very thin, being about 5 kilometers. Such seismic evidence confirms the concept of isostasy: the crust corresponds to the "floating" mass and the mantle to the "fluid" substratum; the light crust made up of granite and basalt is balanced isostatically on the heavier mantle, as shown in Figure 1-13. The mantle, however,

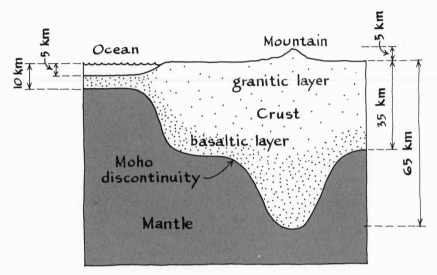

FIGURE 1–13.

Isostatic equilibrium between the crust and the mantle.

is a solid since it transmits S waves. How can a solid behave like a fluid? This question will be treated in Chapter 2 (see page 72).

Thus, the study of earthquake waves has not only revealed the boundaries between the crust, the mantle and the core, but has also confirmed the principle of isostasy. An earthquake, dreaded by many as the worst of calamities, is, nevertheless, an invaluable source of information for the explorers of the earth's interior.

Moving Continents

To return to our main subject: Wegener, in developing his continental drift theory, was encouraged by another geophysical phe-

nomenon, the uplift of the Scandinavian peninsula and the northern
part of North America. Figure 1-14 illustrates the uplift of the
Scandinavian peninsula, still going on today at the rate of about 1
meter per century at the maximum. Such uplift is today measured

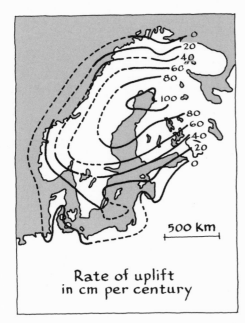

Rate of uplift
in cm per century

FIGURE 1–14.

Contemporary rate of uplift of
Scandinavia. (Adapted from
figure on p. 194 in *Physics of
the Earth's Interior*, by B. Gutenberg (1959), with permission of Academic Press Inc.,
New York.)

by levelling. According to evidences from former shorelines, the
Scandinavian peninsula has been continually rising since the end
of the Great Ice Age about 20 thousand years ago. During the Ice
Age, when this region was covered by thick inland ice sheets (2 ∼ 3
kilometers thick), the peninsula had sunk as much as a few hundred
meters. These phenomena can be interpreted as follows: during the
Ice Age, the Scandinavian peninsula, loaded by glaciers, had sunk
into the mantle; since the ice melted away, Scandinavia has been
steadily rising in accordance with the principle of buoyancy. In
other words, we have here a living evidence of "isostasy."

In the case of isostasy, the movement of continents is vertical
and on a small scale. In the case of continental drift, the movement
is horizontal and may extend over several thousand kilometers. In

spite of these considerable differences, the evidence that continents can move vertically encouraged the supporters of the continental drift theory. If continents move vertically, they argued, why not horizontally?

Transatlantic "Exchange" of Earthworms and Snails?

Another supporting evidence is the distribution of earthworms and garden snails which strongly suggests that there was formerly an exchange of these forms across the Atlantic. In Wegener's day, *helix pomata*, a kind of garden snail, was found only in the western part of Europe and the eastern part of North America. Needless to say, the sea is an insurmountable obstacle to the migration of these creatures. Yet, across the Atlantic, earthworms show a close affinity for the same latitudes. Of particular interest is the *lumbricidae*, a relatively young genus of earthworm which is found in Japan, the Asian and the European continents and, across the Atlantic, on the east but *not on the west* coast of North America. These evidences of faunal distribution strongly suggest that there was formerly a land connection across the Atlantic. Moreover, the split seems to have begun first in the south and gradually opened up northward, for, while the South Atlantic is spanned by the older genera of earthworms, the North Atlantic is spanned by younger ones such as the *lumbricidae*.

Earth science has its humorous side when tiny creatures like earthworms and snails turn out to contribute much to the study of gigantic continents. Evidences of former exchange, however, are not limited to these creatures, but are found in various ancient animals and plants preserved in strata as fossils. Aquatic animals must be ruled out, but when we find, on either side of the Atlantic, fossils of the identical species of a land animal, it is tempting to believe that there was formerly a land connection across the Atlantic. Before Wegener developed his theory, suggesting that some of the continents once directly *adjoined* one another, biological and paleontological evidence for former land connections between continents was explained by the land bridge theory. This theory postulated that the oceans were formerly *bridged* by land masses which subsequently became submerged.

The Land Bridge Theory

Paleontology is a vast and elaborate field of study. The patient efforts of paleontologists as they attempt to unravel the complex process of evolution through study of the fossils deserve our sincere admiration. It is beyond the scope of this book, however, to enumerate and examine all the paleontological evidences advanced for and against the former existence of land connections. We will limit ourselves here to a general synopsis of these evidences compiled by a German scientist, T. Arldt, since that will give us a rough idea of which continents must have been connected when. In this synopsis, since the data are so varied and complex, the following statistical method has been adopted.

Take for example the possible connection between India and Madagascar, the big island near the east coast of Africa, in the Cretaceous period. Some paleontologists provide evidences in favor of a connection, some against it. A favorable evidence is counted as a "plus" and an unfavorable one as a "minus." When all the plus and the minus evidences are counted for each geological period, the history of the possible connection between Madagascar and India comes to light.

This method is of course open to criticism as an oversimplification, but as long as we keep that fully in mind, the result is well worth examining. For a land connection between India and Madagascar, the favorable evidences predominate through all geological periods up to the Cretaceous, towards the end of which the connection seems to disappear. Particularly impressive is the evidence based on the distribution of lemur, a fox-like monkey which is found today in India, Ceylon, Southeast Asia, and, across the Indian Ocean, in Madagascar and some parts of Africa. The imagined land bridge which possibly lay between India and Madagascar across the Indian Ocean was named *Lemuria* after this animal which provides the strongest evidence for the former existence of the bridge. Thus, paleontologists leave no stone unturned, but mobilize every creature that can be useful, from creeping snails to nimble monkeys.

The history of possible land connections between various continents through geological periods is illustrated in Figure 1-15. The

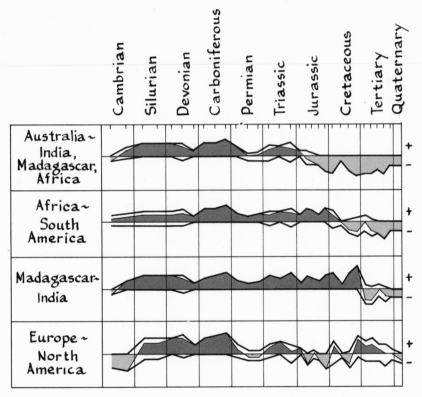

FIGURE 1–15.

History of possible land connections between continents. The number of favorable votes is shown by the upper solid curve, the number of unfavorable votes by the lower solid curve. The difference is shaded in dark gray if favorable, light gray if unfavorable. (From figure 15 in *The Origin of Continents and Oceans*, by Alfred Wegener (1924), with permission of E. P. Dutton & Co., New York and Methuen & Co., London.)

number of favorable and unfavorable evidences are drawn in curves, and the differences between the two are shaded in dark gray or in light gray. In dark gray areas, favorable evidences predominate implying the existence of a land connection. In light gray areas, evidences against its existence predominate.

It is clear at a glance that in geological antiquity, before the Jurassic period, land connection seems to have prevailed between Australia, India, Africa, and Madagascar; between Africa and South America; and between Europe and North America, respectively. Then after the Jurassic (about 150 million years ago), for some reason or other, those land connections seem to have disappeared one after another. Although omitted from this figure, there are evidences for a similar connection between the Antarctica, South America, Australia, and Africa.

No Need for a Land Bridge

As an answer to these undeniable paleontological evidences of land connections, the idea of a land bridge certainly excites the imagination. But is it only a land bridge that makes intercontinental connection possible? According to the continental drift theory, the land bridge is unnecessary since, formerly, the continents themselves directly adjoined one another. This explains very simply why earthworms and snails show such close affinity across the Atlantic. The continental drift theory was originally inspired by the similarity of coastlines, but it was these paleontological evidences, once considered the grounds of the land bridge theory, that gave concreteness to Wegener's theory and enabled him to date the separation of continents mentioned above.

This is perhaps a typical example of how the same evidences can be interpreted with drastically differing conclusions. Formerly, in the light of the firmly held belief that continents are immobile, it was necessary to assume a land bridge to account for the paleontological evidences. Throw away the basic hypothesis, and the land bridge theory becomes unnecessary.

Still, the supporters of the land bridge theory did not cede easily to the continental drift theory. Are there evidences, other than paleontological, which indicate that land bridges existed before the Jurassic period and were later submerged under the ocean?

In the ocean floors of the Atlantic and the Indian Ocean, no indication has been found of the former existence of land bridges. There are mid-oceanic ridges (see page 226, Figure 7-4) rising above the ocean floor, but they generally run parallel with and not cross-wise to the coastline. Moreover, a land bridge extending over several thousand kilometers could hardly have been very narrow relative to its length. Its width would have had to be comparable to its length; in other words, it would have had to have something of the dimension of a continent rather than a bridge. However, as we have already seen, continents and ocean floors are quite different in their crustal structure; the continent has a large amount of granitic material while the ocean floor is made up entirely of basaltic material. No remnant of submerged continental granitic mass has been found on the ocean floor.

But supposing that a "bridge" had actually been submerged by some force; it follows that the sea-water replaced by such a huge bulk would have flooded the greater part of the original continents. Thus, the land bridge theory cannot avoid some self-contradictions. In its place, the continental drift theory offered a new solution to the paleontological enigma as well as to the problem of mountain formation.

A Torn Newspaper

As we have already seen, the continental drift theory was originally inspired by the striking resemblance of the shape of the Atlantic shores. If the two Americas are moved eastward on a globe towards Europe and Africa, the coastlines would roughly fit together, allowing for minor discordances. It is as if we had torn a piece of newspaper into two and joined the parts again at the jagged ends (this should not be attempted with a map on Mercator's projection where, as is well known, the polar regions are unduly stretched).

It is difficult to imagine that such fitting of coastlines is purely accidental, yet obviously more than the apparent fit is necessary to conclude that the Americas were actually joined to the European and African continents.

It is certainly premature to conclude that because two pieces of newspaper fit roughly at torn ends, they must originally have formed a single piece. The fitting of coastlines, so to speak, corresponds to this. But suppose the printed lines, too, fitted across the torn ends and made sense as in Figure 1-16? Then there could

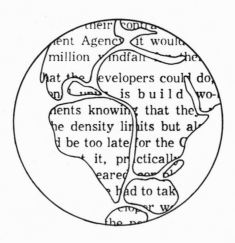

FIGURE 1–16.

Printed lines across the continent.

be no doubt that the pieces came from a single sheet. In the case of continents, it is their geological structures that correspond to the printed lines.

Will the Printed Lines Fit?

If the Atlantic is indeed a rift dividing continents that were formerly joined, then their geological structures should be continuous from one coast to the other; if a mountain range ends on one Atlantic coast, its extension should be found on the opposite coast. According to Wegener, the lines printed on the continents fit excellently.

In the southern end of Africa, there extends from east to west a Permian folded range, the Cape Mountains (see Figure 1-17).

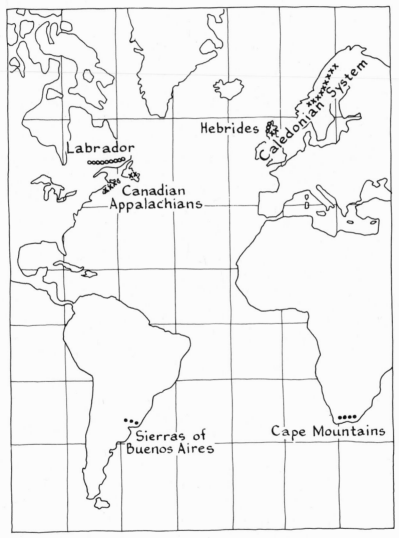

FIGURE 1-17.

Its westward extension can be found in folded ranges strikingly similar in structure and rock composition, lying in the district south of Buenos Aires in South America. The immense gneissic plateau on the African continent also bears close resemblance to that in Brazil (gneiss is a coarse-grained rock with minerals in parallel streaks or bands). Similar agreements are found, moreover, in the distribution of various sedimentary and igneous strata (for explanation of sedimentary, igneous and metamorphic rocks, see pages 156 and 161), and the agreement is particularly impressive in the case of kimberlite (a special kind of rock which contains diamond). We have no space here to enumerate all the examples, but one fact should be noted. These geological agreements disappear completely after the Cretaceous period.

For instance, the Pyrenees in Spain or the Atlas in North Africa formed in the mid-Tertiary period find no extension on the American side. Thus, these geological evidences confirm the inference drawn from paleontological studies that the Atlantic Rift began in the Lower to Middle Cretaceous period.

In the Northern Hemisphere, similar geological agreements are found between northern Europe and North America. Look at Figure 1-17. The Precambrian Hebrides and the gneissic fold mountains in northern Scotland correspond to the Labrador formation on the American side. Among the younger mountain ranges, the Caledonian System in Norway and Scotland, formed in the Silurian-to-Devonian periods, finds its extension in the Canadian Appalachians. Moreover, there are indications from glacial traces that the North Atlantic coasts were still joined in the Quaternary period when the Atlantic Rift had already begun in the south. From such evidences, it was inferred that the Atlantic Rift had first begun in the south and gradually opened up northward.

According to Wegener, if we find one such evidence of geological agreement, the chances are ten to one that the continents were formerly joined. If we find six such evidences, the chances are raised to 10^6 or a million to one. Although this reasoning is questionable, the geological agreements themselves are beyond doubt impressive. Not only across the Atlantic, but between India,

Madagascar, and East Africa too, the gneissic plateaus agree even in their directions of folding, confirming the continental drift theory. For the Southern Hemisphere, however, the best evidence is offered by the glaciers of the Permo-Carboniferous period to which we will now turn our attention.

Glaciers at the Equator

About 300 million years ago, in the Permo-Carboniferous period, the earth was visited by an extensive glaciation. The former existence of glaciers can be easily recognized by the marks of erosion on rock floors and by moraines (debris characteristic of glaciers). Such traces of the Permo-Carboniferous glaciers are found widely scattered over every continent in the Southern Hemisphere. In

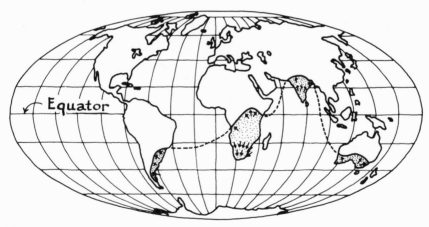

FIGURE 1–18.

Distribution of the Permo-Carboniferous glaciation. The arrows indicate the direction of ice movement. (Adapted from figure on p. 501 in *Principles of Physical Geology*, by A. Holmes (1945), with permission of Ronald Press Co., New York and Thomas Nelson & Sons, London.)

Africa and India, glacial traces are found even near the Equator
(see Figure 1-18). Thus, it is fair to conclude that a considerable
part of the earth's surface must have been under a polar climate.
Yet, in the Northern Hemisphere, no traces of the Permo-Carboniferous glaciers can be found. On the contrary, according to evidences
of fossil plant, tropical climate seems to have prevailed at that
period. How are we to account for such phenomena?

If the continents were once grouped according to Wegener's
theory as in Figure 1-19, the glaciated areas are neatly packed into

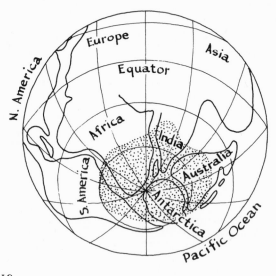

FIGURE 1-19.

Distribution of the Permo-Carboniferous glaciation, with the continents
reassembled according to Wegener's theory of continental drift. (Adapted
from figure on p. 502 in *Principles of Physical Geology*, by A. Holmes (1945),
with permission of Ronald Press Co., New York and Thomas Nelson &
Sons, London.)

a small area. The reasonableness of this explanation speaks for
itself. Wegener's theory found many ardent supporters among
geologists from the Southern Hemisphere, chiefly because they had
seen with their own eyes this mystery of glacial distribution.

Coal in Antarctica

From glaciers at the Equator, we turn to another climatic mystery, coal in Antarctica.

Today, as we all know, the climates on the earth range from the severely frigid near the Poles to the tropical near the Equator. Between these two extremes, the climate is divided into zones such as the subfrigid, the temperate, and the subtropical. These climatic zones are chiefly determined by latitude, and fauna and flora naturally vary from one zone to another. If the same can be said of the climate in the geologic past, we should be able to infer the former latitude of a certain region through fossils of animals and plants discovered in that region. Such study is called paleoclimatology. Some of the data most frequently used in paleoclimatology are the tundra flora (characteristic of the polar climate) and coral (temperate climate). Beside fossils, there are useful data such as glacial traces, and, characteristic of arid climates, rock-salt and certain kinds of sandstone.

A few years ago, the authors had a chance to see a documentary film of an American expedition to the Antarctica. In the film, we saw, as the explorers saw it from an airplane, a vast stretch of coal measures exposed for miles in certain parts of Antarctica. We were struck with wonder—and the wonder is still vivid in our memory— that in this Antarctica, now lying under the severest polar climate and devoid of almost all forms of life, there should once have flourished such abundant plant life, luxuriant enough to have left all this coal. (The coal measures in Antarctica, lying directly on the Permo-Carboniferous moraines, indicate the subfrigid rather than the temperate climate. Nevertheless, the former existence of non-polar plant life itself is indisputable.)

As the above example shows, it can be inferred from paleoclimatological data that, in the past, in most regions of the earth, a climate entirely different from that of today prevailed. Can these paleoclimatological data, like the glaciers at the Equator, help to confirm the continental drift theory? Here, too, the data turned out to be favorable to the theory.

Of course, animals and plants sometimes show an astonishing adaptability and occur in unexpected climatic environments. The climate itself, affected by topography or by ocean currents, does not always fit into simple climatic zones. The temperate climate of northwest Europe, for which the Gulf Stream is responsible, is one such example. Yet, the paleoclimatological data enable us to reconstruct approximately the climatic zones of the past. The result is that the climatic pattern in the past can be best explained if the continents were once grouped according to Wegener's theory. As an example, the climatic pattern in the Carboniferous period, compiled by Wegener and others is shown in Figure 1-20. In this map,

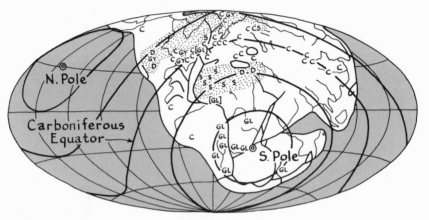

FIGURE 1–20.

Paleoclimatological evidences for the Carboniferous. C, evidence of coal; D, desert sandstone; GL, glacier; GY, gypsum; S, rock-salt; dotted area, arid regions. (Adapted from figure 2 in *Theory of Continental Drift, A Symposium* (1928), with permission of The American Association of Petroleum Geologists, Tulsa.)

C stands for coal measure, GL for glacier, S for rock-salt, GY for gypsum, and D for desert sandstone. The dotted areas represent arid climate. The double circles show the estimated positions of the poles, and the curiously warped concentric circles drawn around the poles represent the parallels at the interval of every 30°.

It is clear at a glance that once the continents are thus grouped, the climatic pattern becomes coherent and easy to explain. Scatter the continents in their present position and the pattern becomes completely baffling. Similar results were obtained for other geologic periods.

Have the Poles Moved?

In the previous section, we tacitly assumed that the position of the poles in the Carboniferous period was considerably different from that of today. Careful readers would have noticed that this requires some explanation. The wandering of poles seems as fantastic an idea as the continental drift theory. Yet paleoclimatologists had suspected and insisted, even before the formulation of the continental drift theory, that the poles have shifted.

Spitsbergen today has a severe polar climate and lies under

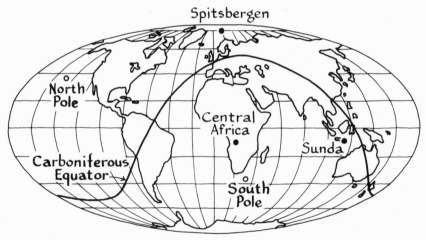

FIGURE 1-21.

Position of the poles and the Equator in the Carboniferous, estimated from paleoclimatology. The reader may trace this equator on a globe himself to prove that this curved line is in fact a circumference of the globe.

land-ice. Yet the fossil record shows that forests of trees of a kind now found only far south in Central Europe still rustled there in the Tertiary period. Further back in the Jurassic period, Spitsbergen was covered with tropical plants. Strangely enough, in Central Africa, which lies 90° south of Spitsbergen (the longitude is approximately the same), exactly the reverse climatic change occurred at about the same period; in the Carboniferous period, the African

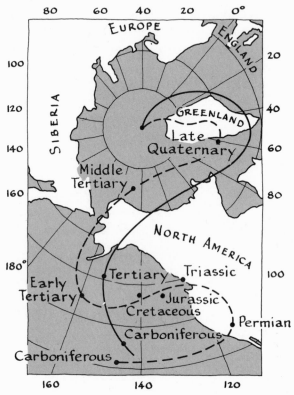

FIGURE 1–22.

Loci of polar displacement according to Wegener (broken curve) and Kreichgauer (solid curve). (Adapted from figure on p. 202 of *Internal Constitution of the Earth*, by B. Gutenberg (1951), with permission of Dover Publication Inc., New York.)

region was covered with land-ice. Moreover, in the Sunda Archipelago lying 90° east of Central Africa (see Figure 1-21), no variation of climate seems to have occurred at all.

How are we to account for these facts? Paleoclimatologists tried to explain them by shifting the poles from their present position and locating them where they would effect a warmer climate in Spitsbergen, a cooler one in Central Africa, and no climatic change in the Sunda Archipelago. Such positioning of the poles and the resulting position of the Equator are shown in Figure 1-21. By similar process, the locus of polar displacement was traced through all geological periods. Some of the results obtained, including that of Wegener, are shown in Figure 1-22. However, the attempts to trace polar displacement through all geological periods always led to contradictions as long as it was assumed that the continents had always been scattered as they are today. For instance, the Permo-Carboniferous glacial traces mentioned earlier are scattered in such a way that any possible position of the pole would leave one or other glaciated area within 20° of the Equator. Because of these contradictions, the theory of polar displacement, in spite of some undeniable evidences for it, was generally distrusted as an absurd idea. For this problem too, the continental drift theory found a happy solution in suggesting that *continents* as well as the poles have drifted. The question of polar wandering, based on modern evidence, will be discussed in Chapter 5 (see page 171).

Modern natural science seeks to keep down to the minimum the number of hypotheses on which arguments are to be based. If arguments come to a dead end, it is proper to reexamine the basic hypotheses and, if necessary, to change them drastically, instead of clinging to old assumptions. Once we do away with the preconception that continents are immobile, we can see how the continental drift theory may cut the Gordian knot and offer solutions to a number of problems which have seemed baffling.

2

Controversy over the Continental Drift Theory

The Identity of Atlas

Greek mythology tells us that the earth is supported by a giant called Atlas. If Atlas moves, so does the earth. Wegener's continental drift theory offers plausible explanations of some of the earth's riddles—the similarity of the shape of the Atlantic shores, the agreement of their geological structures, the origin of some mountains and archipelagos, the paleontological evidence of former land connection between continents, and the mystery of the paleoclimatological pattern—*if* we accept the hypothesis that continents have drifted.

But what force moved the continents? What is the identity of this "Atlas," or, in scientific terms, the *mechanism* of continental drift?

The last chapter of Wegener's work, *Die Entstehung der Kontinente und Ozeane*, is devoted to this problem. Perusal of this chapter, however, leaves the impression that eloquent as Wegener has hitherto been in urging the effectiveness of his own theory as a solution to various concrete problems, the vigor of his argument suddenly flags. To put our conclusion first, Wegener, for all his pains, could not explain how or why, in a convincing manner. We

will now introduce briefly the force which Wegener invoked—
perhaps with a forlorn hope—as the mechanism of continental drift.

The Pole-Fleeing Force

As we saw in Chapter 1, the movement of most continents can be
divided into a westward movement and a movement towards the
Equator. Let us first examine the mechanism of the equatorward
drift according to Wegener.

As we know, every object in the universe attracts every other
object with a force that depends on their masses and on the distance
between them. The earth with its great mass exerts an immense
attraction on objects at its surface. Take an object located at point
O on the earth's surface, in Figure 2-1. The attraction of the earth

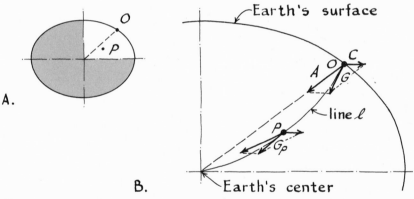

FIGURE 2–1.

A schematic drawing of the directions of the attraction, of the centrifugal
force of rotation, and of the force of gravity. The lengths of the arrows
are not proportional to the magnitudes of the forces.

for this object is represented by arrow A. The attraction is directed
towards the earth's center.

Because the earth rotates on its axis, there is another force, the
centrifugal force of rotation (arrow C), by which the object tends to

be thrown off from the rotating earth's surface, just as mud is thrown from a spinning automobile wheel. The centrifugal force of rotation is greatest at the Equator and steadily diminishes till it becomes zero at the pole. Even near the Equator, the centrifugal force of rotation is about three hundred times smaller than the attraction of the earth; hence solid objects do not fly off the Equator into space. However, the centrifugal force of rotation is by no means insignificant, for it is because of this force that the earth is not quite spherical in shape, but bulges slightly at the Equator and flattens at the poles; the earth, in other words, is an ellipsoid, with the equatorial radius approximately 20 kilometers larger than the polar radius. The shape is drawn, greatly exaggerated, in Figure 2-1.

The object at point O on the earth's surface, then, is subject to both the attraction of the earth (arrow A) and the centrifugal force of rotation (arrow C). The combined effect—the resultant—of the attraction and the centrifugal force of rotation is the *force of gravity*. As the reader knows, the resultant of two forces, A and C, is represented by the diagonal of a parallelogram having A and C for its neighboring sides (see Figure 2-1). Therefore the force of gravity at point O is represented by arrow G in Figure 2-1. Note that arrow G does not point exactly to the earth's center.

The direction of the force of gravity is the direction of the plumb line (explained in Chapter 1, page 31). What direction does the plumb line take in the earth's interior? This cannot be measured directly, but theoretically it is considered as follows. Look at Figure 2-1. The direction of plumb line from point O to the earth's center changes along line l in Figure 2-1. Line l is convex towards the Equator because of the centrifugal force of rotation. The direction of the plumb line at point P on line l is given by arrow G_P which is tangent to line l at P (Figure 2-1). Note that the direction of G_P differs slightly from that of G. There is always some difference in the direction of plumb line at any two points on line l.

The force of gravity exerted on an object can be considered as a single force acting at its *center of gravity*. The center of gravity is the specific point within the object at which the force of gravity acts no matter how the object is oriented. For instance, in a right cylinder of uniform density, the center of gravity is its geometric center P, whether the cylinder is lying horizontally (Figure 2-2A) or tipped sidewise (Figure 2-2B).

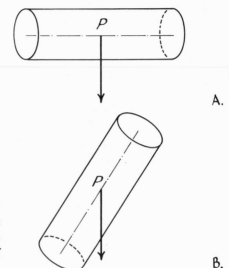

A.

FIGURE 2–2.

The center of gravity of a right cylinder of uniform density. The arrows indicate the directions of the force of gravity.

B.

According to the idea of isostasy, the continental mass floats on the mantle in accordance with the principle of buoyancy (see page 33). Let us examine what forces are at work in a floating continent. Look at Figure 2-3. The force of gravity (arrow g) acts in the direction of the plumb line at the center of gravity G of the continental mass. The buoyancy (arrow b), on the other hand, acts upward, in a direction opposite to that of the plumb line, at the center of buoyancy B, which by definition is the center of gravity of the part of the continental mass submerged in the mantle (the shaded area). As can be seen in Figure 2-3, the center of buoyancy B is usually lower than the center of gravity G, and the force of gravity g and buoyancy b offset one another. However, as we have just seen, the direction of plumb line at point G differs slightly from that at point B. Consequently, g and b do not offset each other completely, but leave a small force P. P is directed toward

Sliding Continents

Geologists who tend to be realistic, also generally considered that Wegener's idea was too fantastic. But it found a few adherents among those who either could not dismiss so lightly the idea of the horizontal movement of continents, or who felt very keenly that the paleontological pattern can only be explained by the former union of continents.

Independently of and earlier than Wegener, in 1910, an American geologist, F. B. Taylor, had suggested that continents have drifted towards the Equator on a large scale (*19*). His motive was to explain the origin of Tertiary mountain ranges such as the Himalayas, the Andes, and the Alps. Taylor was impressed by the fact

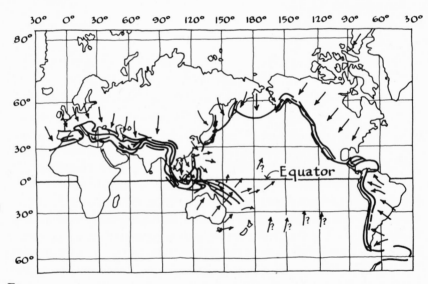

FIGURE 2–4.

Principal Tertiary mountain ranges of the world. The arrows show the general direction of crustal movement in the continents. (Adapted from figure 27 in Sliding continents and tidal forces, by F. B. Taylor, in *Continental Drift, A Symposium* (1928), with permission of the American Association of Petroleum Geologists, Tulsa.)

that the Tertiary ranges on the Asian Continent are mostly arc-shaped, curving out southward (see Figure 2-4). From this fact, he inferred that these mountains must have been produced by a southward sliding of the continent. As we mentioned earlier (see page 59), the centrifugal force due to the earth's rotation leaves a small equatorward force. Taylor assumed that, because of this force, continents in the Northern Hemisphere slid southward and those in the Southern Hemisphere northward (see Figure 2-4).

How does a continental slide produce arcuate (arc-shaped) mountain ranges? Taylor's idea is illustrated in Figure 2-5. Let

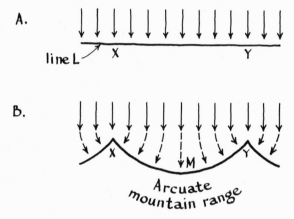

FIGURE 2–5.

Formation of arcuate mountain ranges according to Taylor. (Adapted from figure 26 in Sliding continents and tidal forces, by F. B. Taylor, in *Continental Drift, A Symposium* (1928), with permission of the American Association of Petroleum Geologists, Tulsa.)

the straight line L represent the front of the crustal sheet before it began to move (Figure 2-5A). At first, the crust moves evenly along the whole line, but if it encounters, at X and Y, slight obstacles which check the movement, the crustal movement would be more vigorous in the middle part (M) than at the sides and an arcuate lobe would be formed (Figure 2-5B). As the crustal sheet spreads out in this form, the moving edge eventually encounters the resistance of the mass ahead, and is folded into mountain ranges.

The mass of the Indian peninsula was a great resisting obstacle in the way of the southward crustal movement, and consequently the Himalayas were piled up. Moreover, the force of movement, checked by the Indian peninsula, was partly deflected eastward into Indochina (the peninsula that now comprises South Vietnam, North Vietnam, Cambodia, Laos, Thailand, Malaya, and Burma) causing a great southward slide in these areas. This is analogous to a stream of water, which, meeting an obstacle that fills half the width of its channel, is retarded and slightly raised against the obstacle, only to rush with accelerated speed through the remaining space.

Taylor's theory had some plausible points and won favor with some scientists. But his theory had one fatal weakness, and once again this was the question of mechanism, the inevitable Atlas.

Did the Earth Capture the Moon?

According to Taylor, the continental slide, caused by the small equatorward force due to the earth's rotation, occurred in the Tertiary period. But why in the Tertiary which is a mere yesterday in the earth's history (see Table 1-1, page 24)? The logical consequence of Taylor's theory would be that the earth's rotation began, or else increased its speed, in the Tertiary. To support this argument, Taylor put forth an extraordinary hypothesis that, towards the end of the Cretaceous or the beginning of the Tertiary, the earth captured as its satellite the fast-rotating moon, and that consequently the speed of the earth's rotation was increased. If we are to believe this, the earth did not have its satellite until the end of the Cretaceous period. The idea is certainly fantastic, but when we come to think of it, there is no positive evidence denying it. In geology, there are theories that are extremely difficult to verify. That, in one sense, makes geology all the more exciting. But in this case, since the idea of continental slide is in itself a hypothesis difficult to verify, it is unsound to base this hypothesis on a still more dubious one, namely, the capture of the moon.

Taylor's theory has another grave weakness in that the formation of mountain ranges is not confined to the Tertiary period. Older mountain ranges, folded earlier in the earth's history, exist. How are we to account for them? According to Taylor's thesis, the earth would need many more moons.

Two Primordial Continents

Taylor assumed, in order to explain the origin of the Himalayas and the Alps, that the continents in the Northern Hemisphere slid southward towards the Equator. In fact, some scientists have suggested that, from the middle of the Paleozoic era to the Tertiary period, the southern part of the Asian Continent (including the Himalayas) and the Mediterranean region (including the Alps) were submerged under a huge geosynclinical ocean named the "Tethys" or the "Tethyan geosyncline" (for explanation of geosyncline, see below). Thus, instead of assuming *one* primordial continent (the Pangea) like Wegener's, many scientists were of the opinion that there were *two*, one in the Northern and the other in the Southern Hemisphere. The leading proponent of this idea was A. L. Du Toit, a professor at Johannesburg University in South Africa. An enthusiastic supporter of the continental drift theory, Du Toit dedicated his major work, *Our Wandering Continents*, to Wegener, with these words: "To the memory of Alfred Wegener for his distinguished services in connection with the geological interpretation of our earth" (*4*).

Deeply as he admired Wegener, Du Toit differed from him in assuming *two* primordial continents separated by the Tethyan geosyncline. Look at Figure 2-6. The dotted areas in the north, representing parts of Europe, Asia, North America, and Greenland were once grouped in a single continent "Laurasia," and those in the south, representing parts of South America, India, Australia, Africa, and the Antarctica, formed the "Gondwana." Like Wegener, Du Toit devoted the last chapter of his book to the

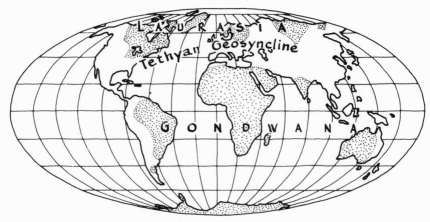

FIGURE 2–6.

The Tethyan geosyncline and the two primordial continents. The dotted areas represent parts of the continents that were formerly grouped as Laurasia in the Northern Hemisphere and Gondwana in the Southern Hemisphere. (Adapted from figure 6 in *Our Wandering Continents*, by A. L. Du Toit (1937), with permission of Oliver and Boyd Ltd., Edinburgh.)

mechanism of continental drift. At a time when several explanations had already been rejected as unconvincing, Du Toit's appeared as a new and original idea.

Continental Ship Equipped with an Engine

To understand Du Toit's idea, we first need to know about a geosyncline. The concept of geosyncline is today generally accepted. Look at Figure 2-7. At the margin of a continent, sediments would accumulate because of erosion and deposition and would form a thick pile of strata. The thickening strata sink under their own weight at about the same rate that the sediments accumulate with the result that the margins of continents are always on lower levels than inland plateaus. Strata thus piled up sometimes attain

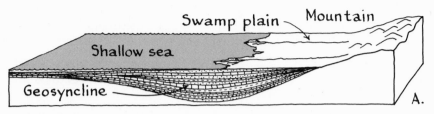

FIGURE 2–7.

Schematic drawing of a geosyncline. (Adapted from figure 297 in *Physical Geology*, by Longwell, Knopf, and Flint (1949), with permission of John Wiley & Sons, New York.)

the thickness of 10 kilometers, and the trough filled with such sediments is called a *geosyncline*.

Figure 2-8 illustrates Du Toit's idea. The continent is considered as a huge mass. At the continental margin (A), a geosyncline forms and slowly sinks. With the sinking of the geosyncline, the primordial continent becomes slightly tilted (Figure 2-8) and begins to slide towards the ocean. Because of this slide, the middle part of the continent becomes subject to tension and tends to split apart. Molten *magma* (rock melted deep within the earth) welling up through the fissure, helps to split the continent farther apart and

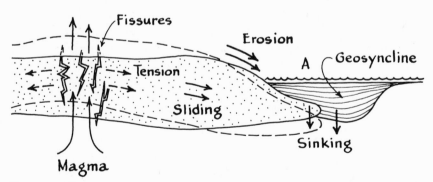

FIGURE 2–8.

Mechanism of continental drift according to Du Toit.

pushes up this middle part so that the split continent is propelled forward and slides further outward.

Thus, by making use of all these geological phenomena, namely, erosion, deposition, and depression of the geosyncline resulting from gravitational energy, and ascent of molten magma resulting from thermal energy, the continent works a kind of engine and "sails" over the mantle. Such, according to Du Toit, is the mechanism of continental drift. In the navigation of this "continental ship," the traditional forces—the tidal force and the pole-fleeing force—have secondary effects and possibly determine the direction of the drift. Unfortunately, Du Toit's theory, original as it was, lacked quantitative evidences and sounded too much like a clever invention. For this reason, it has been more or less forgotten since his days.

The Balance of Pros and Cons

In 1928, a symposium on the continental drift theory was held in New York. This symposium, sponsored by the American Association of Petroleum Geologists, brought together many leading geologists of the time and marked an epoch in the history of the continental drift theory.

Among the fourteen geologists who gave their views at the symposium, five actively supported the continental drift theory, two supported it with some reservations, and the remaining seven opposed it. Thus the pros and cons were roughly half and half. What were the chief points of dispute at the symposium?

Objections to the Drift Theory: 1

Among the opponents were, as noted, some eminent scientists, and their attack was vigorous. One distinguished geologist declared:

> After considering the theory of continental drift with avowed impartiality, I conclude by means of geophysical, geological and palaeontologic reasoning that it should be rejected. (19)

What were the chief grounds for such rejection? In the following discussions, some old technical terms used in Wegener's day but no longer prevalent, are replaced by their modern equivalents.

To begin with, there is the problem of "Atlas." What force could have dragged the gigantic continents over the mantle? The forces invoked by Wegener and his supporters are far too small and hopelessly inadequate. This difficulty, as we have already seen, was admitted by the adherents themselves.

The second point of objection concerns the rigidity of the crust and the mantle. According to Wegener, the heavy rocks which compose the mantle are apparently solid and yet, given enough time, they begin to flow and yield to the crustal mass. Thus it was possible for the continental mass to make its way and drift through the mantle. According to this reasoning, the crust is more rigid than the mantle. Yet elsewhere, Wegener asserts that mountain ranges were formed on the front margin of drifting continents when the continental mass encountered the resistance of the mantle and was compressed into fold mountains. Then does it not follow that the crust is weaker than the mantle? How can the crust be at once more rigid and weaker than the mantle? Can we thrust a leaden chisel into steel? Does a steel nail bend when it is driven into cream cheese?

Let us go on to the third point. The earth has a history of several billion years. The post-Cambrian fossiliferous age alone encompasses several hundred million years. Yet, according to Wegener, the primordial continent did not begin to split until the Mesozoic era. Is it not strange that a great incident like the continental drift should have been confined to a very recent period in the earth's history? Why did it not occur before? This argument, like the objection to Taylor's theory, is based on the idea that the earth was far more active in remote antiquity than it is now. But quite apart from this notion, it does certainly seem strange that such a change should have taken place only very recently and in such a short interval of time (only about one-sixtieth or one-seventieth of the total history of the earth).

As we shall see later, the adherents of the continental drift theory did their best to refute such objections. When theoretical difficulties could not be overcome, they argued that facts should not be ignored just because there are no theoretical explanations for them. No one could question the fact that thick glaciers covered the Northern Hemisphere in the Quaternary period, although theoretical explanation is lacking. It is an undeniable fact that intensely folded mountains such as the Alps exist, but for them no satisfactory explanation (except the continental drift theory) has been found.

Objections: 2

Are the facts indicated by Wegener undeniable then? Objections were raised against them by a number of well-known geologists.

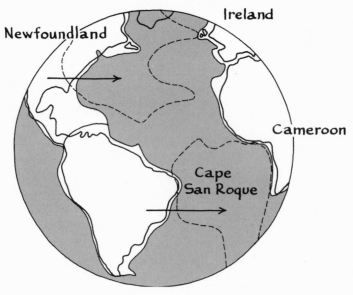

FIGURE 2–9.

Matching of Atlantic coastlines. (Adapted from figure 15 on page 110 of *Continental Drift, A Symposium* (1928), with permission of the American Association of Petroleum Geologists, Tulsa.)

To begin with, they questioned the similarity of the Atlantic shores, the very starting-point of Wegener's theory. Look at Figure 2-9 which shows the result of an experiment on continental drift. A transparent cover is fitted tightly over the globe and the shapes of the continents (including the continental shelves) are traced on the cover. The North America on the cover is moved eastward according to Wegener's instruction so that Newfoundland is placed beside Ireland. Similarly, South America is moved eastward so that Cape San Roque of Brazil is placed beside Cameroon in Africa. Although some parts fit perfectly, Central America is about 2000 kilometers away from Africa, leaving the Atlantic still as a great ocean. One opponent declared:

> It is evident therefore that Wegener has taken extraordinary liberties with the earth's rigid crust, making it pliable so as to stretch the two Americas from north to south about 3000 kilometers.(19)

As for the fitting of the torn newspaper, it was found that the strata in Ireland and Newfoundland show some geological similarity, but are far from a perfect fit. Paleontological evidences too were disputed; it was pointed out that only 5% of faunal species instead of the 50 ~ 75% claimed by Wegener, show positive affinity across the Atlantic.

When arguments are thus specialized, it is all but impossible for a layman to judge which argument is reliable. Each scientist can more or less draw arbitrary conclusions, depending on his basic experience and attitude. One opponent went so far as to say that Wegener's method:

> . . . is not scientific, but takes the familiar course of an initial idea, a selective search through the literature for corroborative evidence, ignoring most of the facts that are opposed to the idea, and ending in a state of auto-intoxication in which the subjective idea comes to be considered as an objective fact.(19)

Beside the objection that the Atlantic coastlines do not fit well, there was another argument. It would have been strange if the continents *had* retained their original shape after a large-scale displacement, which would surely have distorted and impaired them. Even if it were shown that the coastlines and strata happened to match perfectly, it would disprove rather than prove the continental drift theory. If indeed, both agreement and disagreement of coastlines would tell against it, the continental drift theory certainly would stand no chance at all of general acceptance.

The Chairman's Concluding Remarks

The symposium was concluded by the Chairman's "Remarks on papers offered to the symposium." The Chairman, W. A. J. M. van Waterschoot van der Gracht of Holland, admitted that the mechanism of the drift was inadequately accounted for, but argued that as a solution to the enigma of the paleontological distribution, the continental drift theory was better than the land-bridge theory (page 42). Adequate mechanism for the drift had not been found up to the present (1928), but might well be found in the future.

On the question of the agreement of the Atlantic shores, the Chairman accepted various points of view and concluded that, although we cannot expect a perfect fit after large-scale displacement, *general similarity* of shape should be, and *is*, in fact, preserved. As for the objection that Wegener's continental drift was confined to too recent a period in the earth's history, the Chairman pointed out that Wegener discussed the history of continents as far as it could be inferred from actual evidences and claimed nothing about their earlier history, concerning which we have such scarce data. In remote antiquity, there might or might not have been continental drift on a large scale. It was unreasonable to attack Wegener on the arbitrary assumption that there was no earlier displacement.

The Chairman's concluding remarks, although inconclusive on some points, were on the whole reasonable.

Hard Fluid and Soft Solid

On the question of the rigidity of the crust and the mantle, the Chairman referred to the following well-known example drawn on by Wegener.

Sealing-wax and pitch are apparently very hard at normal temperature, solidly brittle enough to break when dropped on the floor. A piece of beeswax is apparently softer, and we cannot immediately push it into a block of sealing-wax. Yet if we leave a piece of beeswax on top of sealing-wax for a long time (see Figure 2-10A),

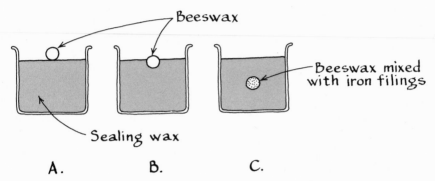

FIGURE 2-10.

the former will gradually sink into the harder material. As beeswax has a lower density than sealing-wax, it stops sinking when isostatic balance is attained (see Figure 2-10B). If we mix some iron filings with the beeswax to give it a higher density than sealing-wax, the piece will gradually sink completely into the sealing-wax (Figure 2-10C). Thus, in the long run, the sealing-wax yields to the small but steady weight of beeswax.

A figure made of hard sealing-wax or pitch will not hold its shape but will gradually yield to the steady force of gravity and become a shapeless mass. Figures made of soft beeswax will endure, like the famous Madame Tussaud's wax figures in London. Thus, material like pitch or sealing-wax behaves like a solid under short-term stresses, but under long-enduring stresses—even if the stresses are small—begins to flow or yield. Such material is called a viscous fluid. On the other hand, material like beeswax is soft under large stresses but will not start flowing under stresses—however long-enduring—unless the stresses applied exceed a certain limit or force. Such material is called a soft solid. Such behavior of material is today studied in rheology (the study of the flow of matter), and sometimes presents strange phenomena which are contrary to common experience and difficult to grasp, yet ultimately are seen to be reasonable.

In the continental drift theory, the earth's crust is considered to be a soft solid and the mantle a viscous fluid. According to Wegener, the evidence of isostatic adjustment such as the uplift of the Scandinavian Peninsula (see page 40) strongly supports this idea, for when the crustal mass is uplifted, the underlying material must flow in to fill the gap; the mantle, then, must behave like a viscous fluid.

How, then, were mountains folded on the front margin of a moving continent? In Wegener's day, the existence of the oceanic crust was not known and the ocean floor was believed by scientists to be made up of mantle substance. The mantle, while behaving like a fluid under the high temperature at great depth, is rigid under the low temperature that prevails at the shallower depth near the surface of the ocean floor. Therefore, according to Wegener, the moving continents may have encountered a considerable resistance from the ocean floor. The crustal mass, being a soft solid, yields to stresses when they exceed a certain limit. Therefore, it is possible that the advancing edges of continents, meeting the resistance of the ocean floor, were compressed and folded into mountain ranges.

The overall conclusion at the symposium was this: the theory of continental drift should not be rejected lightly, for if there are grave objections to it, there are also sound supporting evidences and some of the objections are refutable. Certainly, the theory was worth much consideration and study, since potentially, if it could be proved, it would provide answers to many questions we have had about the earth.

History Repeats Itself

One of the strange phenomena that geologists have found about the earth's history is that geologic revolutions occur periodically. Technically, a geologic revolution is called an orogenesis. During an orogenesis, the earth's crust becomes universally active, accompanied by volcanic eruptions, metamorphism (see page 161) of rocks and formation of great mountain ranges such as the Alps and the Himalayas. As far as geologists have been able to trace back the earth's history, such orogeneses have occurred several times in the past (see Table 1-1, page 25). An orogenesis lasts for about a hundred million years followed by a long period of quiescence. The lofty peaks that we know today, the Alps, the Himalayas, and the Andes for example, were born during the most recent of such revolutions, the Alpine orogenesis, which began in the late Triassic and culminated in the middle Tertiary period.

The periodic occurrence of orogeneses has long been a puzzle to geoscientists. If the earth is really cooling at a steady rate, why should orogeneses occur only periodically? An Irish scientist, John Joly, studied this problem and worked out a new theory between 1923 and 1926. At the New York Symposium mentioned above, Joly's theory gave a ray of hope to the supporters of the continental drift theory. Before going into his theory, let us look at some facts about radioactivity, since it forms the background of Joly's theory.

Radioactivity Heats Up the Earth

The traditional belief about the earth's history was that the earth had originally been a great ball of hot gas which gradually cooled, and condensed into the sphere of liquids and solids that we call the earth. Because the earth is hotter in its interior than it is on the surface—as volcanic eruption of melted rock shows—people imagined that the earth had been hotter in the past and has subsequently cooled.

In 1896, however, Becquerel discovered radioactivity. This discovery opened new vistas not only in physics but in geology as well; our image of the earth was drastically changed, because a source of heat, radioactive elements, was found to exist in the earth's interior. A descriptive explanation of radioactivity may be superfluous to readers in this atomic age. However, this phenomenon of nature is so significant in the study of the earth that we will digress briefly from our main theme to describe the process of radioactivity, especially those of its aspects that make it function as the earth's furnace.

By 1958, 102 elements had been discovered. An *element* is a substance that consists of atoms of only one kind. An *atom* has a nucleus made up of one or more protons and, except in ordinary hydrogen, one or more neutrons. The nucleus is surrounded by a "cloud" made up of one or more rapidly moving electrons. This is shown schematically in Figure 2-11.

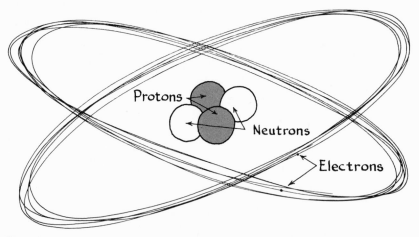

FIGURE 2–11.

Schematic drawing of the atom of helium.

The electron, the proton and the neutron are the three fundamental particles that make up an atom. The electron has one unit of negative electrical charge and the proton one unit of positive electrical charge. The neutron, as the name implies, is electrically neutral.

In any atom, protons and electrons are always present in amounts that add up to electrical neutrality—that is, there is always the same number of protons as of electrons. However, an atom may gain or lose one or more electrons, by chemical reaction for instance, and become negatively or positively charged. Such a charged atom is called an *ion*.

The number of protons in the atom, called the *atomic number*, determines the chemical characteristics of the atom. Atoms range from the simplest kind, the hydrogen that has only one proton in its nucleus, to nobelium that has 102.

The mass of a proton or a neutron is of an amount which we term "one unit of mass." By contrast, the electron has a mass so much smaller (about two thousand times smaller than that of a proton) as to be negligible. As the mass of an electron is negligible, the mass of an atom can be expressed by the sum of the number of protons and the number of neutrons in the nucleus. The sum is called the *mass number* of an atom. For example, every atom of uranium has 92 protons but some atoms have 143 neutrons and others have 146. In the respective cases, the mass numbers of these otherwise similar atoms of uranium are 235 and 238. These variants—the atoms that have the same atomic number but different mass numbers—are called *isotopes*.

Of the 102 elements, some naturally emit "radioactive rays." These rays are chiefly of three kinds, alpha, beta, and gamma rays. Alpha rays consist of the equivalent of the atomic nuclei of the element helium (see Figure 2-11). Beta rays consist of "streams" of electrons emitted at high speed (the speed depends on various conditions, but sometimes it approximates that of light which is 3×10^{10} cm/sec). Gamma rays are streams of what are called electromagnetic waves. Light rays and radio waves that we know well from everyday experience are also electromagnetic waves, but they differ from gamma rays in wavelength; the wavelengths of gamma rays are of the order of 10^{-8} to 10^{-9} cm, those of visible light rays $4 \sim 7.7 \times 10^{-5}$ cm and those of radio waves used for communication of the order of 10^{-1} to 10^{6} cm.

The nuclei of atoms of radioactive elements are unstable and decompose spontaneously. This *spontaneous disintegration* consists of the emission of alpha and beta particles which changes the nuclear structure of the element and thus transforms it into a different element. For instance, U^{238} emits alpha rays and becomes the element

Th²³⁴. This is shown schematically in Figure 2-12. For convenience, the original radioactive element, here U²³⁸, is called a *parent* element; the product of the emission of rays, here Th²³⁴, is called a

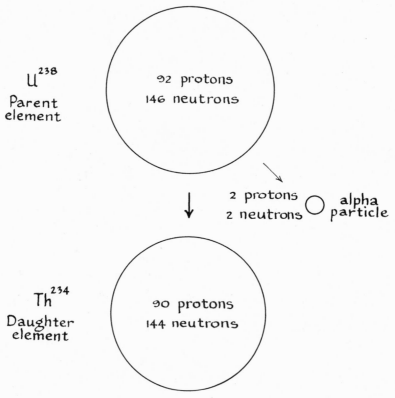

FIGURE 2–12.

Disintegration of U²³⁸ into Th²³⁴ by emission of alpha rays.

daughter element. We call such an element a "daughter" rather than a "son" because a daughter can do what a son cannot; she may herself have offspring. Table 2-1 lists all the daughter elements of U²³⁸. The ultimate product of a radioactive element is a barren daughter, one that emits no rays; in the case of U²³⁸ the ultimate daughter element is lead, Pb²⁰⁶.

The emission of radioactive rays is accompanied by a release of heat. The amount of heat released by the disintegration of one

TABLE 2-1 *

THE URANIUM (U^{238}) SERIES

Isotope	Particle emitted	
$_{92}U^{238}$	α	parent element
$_{90}Th^{234}$	β	daughter element
$_{91}Pa^{234}$	β	daughter element
$_{92}U^{234}$	α	daughter element
$_{90}Th^{230}$	α	daughter element
$_{88}Ra^{226}$	α	daughter element
$_{86}Rn^{222}$	α	daughter element
$_{84}Po^{218}$	α	daughter element
$_{82}Pb^{214}$	β	daughter element
$_{83}Bi^{214}$	$\begin{cases} \alpha\ 0.04\% \\ \beta\ 99.96\% \end{cases}$	daughter element
$_{84}Po^{214}$	α	daughter element
$_{81}Th^{210}$	β	daughter element
$_{82}Pb^{210}$	β	daughter element
$_{83}Bi^{210}$	β	daughter element
$_{84}Po^{210}$	α	daughter element
$_{82}Pb^{206}$		stable daughter element

* Based on Table 8-3 in *Physics and Geology*, by Jacobs, Russell, and Wilson (1959), with permission of McGraw-Hill Co., New York.

atom of U^{238} to Pb^{206} is 1.85×10^{-12} calorie. If we waited long enough for one gram of uranium to disintegrate to stable lead, the amount of energy liberated would be equivalent to that of burning 800 kilograms of coal.

The rate of spontaneous disintegration varies tremendously with different elements. It is expressed in terms of the element's *half-life*, which is the time required for half its atoms to disintegrate.

Radium (one of the daughter elements of U^{238}, see Table 2-1) has a half-life of 1622 years. If we start with 10 grams of radium, in 1622 years, 5 grams will be left; in another 1622 years, only 2.5 grams will remain and so on. As far as scientists have been able to test by experiments, the rate of disintegration is never changed by temperature, pressure, or the state of chemical combination of the element. The half-life of a radioactive element is thus considered a material constant, that is, a fundamental property of the element.

Some radioactive elements have immensely long half-lives; U^{238} has a half-life of 4.5 billion years and Th^{232} 13.9 billion years. Thus the half-life of the latter actually outspans the earth's history itself. By contrast, some are extremely short-lived; the tenth daughter element of U^{238}, Po^{214} (see Table 2-1) has a half-life of about one-millionth of a second. Parenthetically, it is interesting to note that radioactive elements with long half-lives proved to be useful clocks of the earth. This will be discussed in some detail in Chapter 6 (see page 211).

Since the rate of disintegration is constant for each element, it follows that the rate of heat release which accompanies disintegration is also constant. Radium, for instance, releases heat at the constant rate of about 140 calories per hour per gram of radium. While the *rate* at which an element emits this heat energy is unchanging, the *total amount emitted* does decrease as the amount of remaining element decreases; a piece of radium weighing 10 grams would release 1400 cal/hr (heat which could melt an ice cube weighing 17.5g), but in 1622 years only 5 grams of radium would be left, releasing only 700 cal/hr, although the rate of heat release per gram of radium remains unchanged.

Radium was isolated from a naturally occurring mineral called pitchblende by Marie Curie soon after the turn of the present century. Pitchblende is an oxide of uranium, dark green to black in color. It also contains lead and very small amounts of other elements, including radium. In the pitchblende used by Curie, radium constituted about one part to 150 millions of pitchblende by weight.

The discovery of radioactivity and of the heat energy that

might result from it spurred scientists to study the question of sources of heat in the earth. The results to date have been far-reaching in our understanding of nature and the earth, and revolutionary for scientific thinking. They are of especial interest to us in considering the theory of continental drift.

Specifically, scientists have found that the chief sources of radioactive heating in the earth are uranium (U), thorium (Th) and an isotope of potassium (K). The most common isotope of potassium, the atomic number of which is 19, has a mass number of 39 because it has 20 neutrons, and this isotope is not radioactive. The isotope with 21 neutrons, which gives this isotope a mass number of 40, is radioactive.

Table 2-2 shows the amount of uranium, thorium, and K^{40}

TABLE 2-2

RADIOACTIVE CONTENT AND HEAT RELEASE

Half-life (yrs.)	Amount contained in rock (gm/ton)			Quantity of heat (cal/gm sec $\times 10^{-16}$)				Total quantity of heat (cal/cm³ year $\times 10^{-5}$)
	U 4.5 billion	Th 13.9 billion	K^{40} 1.3 billion	U	Th	K^{40}	Total	
Granitic rock......	4	13	4.1	940	820	300	2060	1.74
Basaltic rock......	0.6	2	1.5	140	130	110	380	0.35
Peridotite....	0.02	0.06	0.02	4.7	3.7	1.5	10	0.01
Chondritic meteorite..	0.011	?	0.093	3	?	7	10	0.0095
Iron meteorite..	0.0001 ~ 0.000001	?	?	0.02 ~ 0.0002	?	?	0.02 ~ 0.0002	0.00006 ~ 0.0000006

contained in rocks. The unit of measurement is one part to a million; that is, where the table indicates "1" there is one gram of radioactive material in one ton of rock. As can be seen from this table, the quantity of radioactive elements is relatively small.

This is fortunate because otherwise the earth might have been un-inhabitable by living species as we know them.

It is clear from Table 2-2 that radioactive elements are most abundant in the granitic rock, less abundant in the basaltic rock and least abundant in peridotite. The granite, the basalt, and the peridotite are good candidates for consideration as the chief compo-nent of the upper crustal layer, the lower crustal layer, and the mantle, respectively. Therefore, we may conclude that radioactive elements, which supply heat energy to the earth, are closely con-centrated near the earth's surface. (The amount of radioactive el-ements in the core of the earth is considered to be comparable to that in meteoritic iron, which, as the Table shows, is extremely small.)

How much heat do these radioactive elements in the rocks generate? Look at the figures in the extreme right column of Table 2-2, under the heading *Total quantity of heat*. Each figure represents the quantity of heat (in calories) generated in 1 cm^3 of rock per year. The quantity is extremely small, so small that it would take 500 million years to boil a glass of water with radioactive heat generated in 1 cm^3 of granite, if the glass is completely insulated. Small as it is, the same quantity of heat would suffice to bring 1 cm^3 of granite to the melting-point in several billion years if there is no escape of heat from the surface. Thus, radioactive heat amounts to very little in terms of our human time-scale, but in terms of geologic time-scale it is by no means so insignificant. With this preliminary knowledge, let us now turn to Joly's theory (*11*).

Joly's Theory: History Repeats Itself

As we noted earlier, great geologic revolutions occurred periodically in the earth's history. An orogenesis begins by the depression of the continental mass and the transgression of the sea over its mar-gin. This, in turn, leads to erosion and deposition, until thick piles of strata—geosynclines—are formed on the continental margin. After a period, these strata are compressed, folded, and finally uplifted into great mountain ranges. There are evidences that mountains such as the Appalachians were formed by the uplift of such a geosynclinal pile: not only are their strata much thicker than the strata of the same age on the adjacent plain, but also the fossils they contain are nearly all of shallow-water organisms. The cause of uplift, however, is still a controversial problem. One view is illustrated in Figure 2-13. The thick strata of a geosyncline (Figure 2-13A) are compressed and folded by horizontal compression (Figure 2-13B). Erosion wears away the elevated part (Figure 2-13C). Because of lessened load, the geosyncline is uplifted slowly in accordance with the principle of isostasy (Figure 2-13D). Joly, as we shall see, suggested another possible cause for uplift.

An orogenesis is also characterized by igneous activities, such as the ascent of magma. Such an orogenesis recurs after about 100 million years of quiescence.

What Joly suggested was that the periodicity of orogeneses is due to radioactivity within the earth (*11*). The basis of this idea is that, because of the low thermal conductivity of the crustal material, radioactive heat generated within the earth accumulates below the crust and gradually melts the subcrustal material (the mantle).

Temperature within the earth increases with depth (see Chapter 6). In other words, heat is flowing from the earth's interior outward. The quantity of heat which escapes into air at the earth's surface is called terrestrial heat flow. This can actually be estimated from the measurement of the downward temperature increase and the thermal conductivity of rocks near the earth's surface. (Thermal conductivity indicates how effectively heat is transferred in a certain material; metal, for instance, has a high thermal conductivity; cork and timber have low thermal conductivities.) According to such measurements, the terrestrial heat flow is estimated at 1.5×10^{-6} cal/cm^2 per second, in other words, $30 \sim 40$ cal/cm^2 per year. Now, according to Table 2-2, the quantities of heat generated by granitic and basaltic rocks are 1.74×10^{-5} cal/cm^3 and 0.35×10^{-5} cal/cm^3 per year respectively. A simple calculation shows that, if the crust consists of 15 km of granite resting on 15 km of basalt, the quantity of heat generated in a square pillar in the crust whose height is 30 km and whose base area is 1 cm^2 would amount to 31

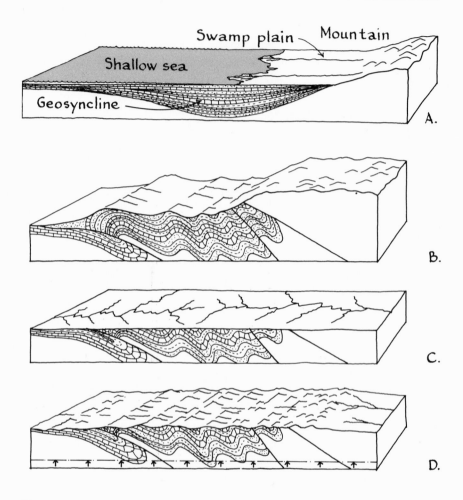

FIGURE 2–13.

A possible cause for the uplift of a geosyncline. (A) The geosyncline re-
ceiving sediments. (B) The geosyncline folded by horizontal compression.
(C) Erosion removes the elevated part. (D) Because of lessened load, the
geosyncline is uplifted isostatically. The broken line corresponds to the
base of the block in (C). (Adapted from figures 297, 298, 299 and 300 in
Physical Geology, by Longwell, Knopf, and Flint (1949), with permission
of John Wiley & Sons, New York.)

calories per year. As we have just seen, the quantity of heat escaping at the earth's surface is 30 $\sim$ 40 cal/cm² per year according to actual measurement. A comparison of these two figures shows that the terrestrial heat flow escaping from an area of 1 cm² at the earth's surface is roughly equivalent to the quantity of heat generated within the crust underlying that area. What happens then to the heat generated *below* the crust? Such heat, unable to escape, accumulates under the crust and melts the mantle. This was the starting-point of Joly's theory.

What happens when the mantle begins to melt? Melting rocks generally expand in volume and become lighter. When the medium of floatage, the mantle, thus becomes lighter, the crust, which has been floating isostatically on the mantle, sinks a little. This, according to Joly, is the cause of the depression of the continental mass at the beginning of the orogenesis. According to his calculation, the continental mass sinks as much as 1.4 kilometers.

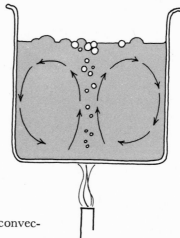

FIGURE 2–14.

A familiar example of thermal convection.

Now, when the mantle is partly melted by radioactive heat, thermal convection occurs in the molten portion. Look at Figure 2-14 which illustrates a familiar example of thermal convection. The heated portion at the bottom expands, becomes light, and floats up to the surface where it is cooled, becomes heavy, and goes down again. This circulation ensures effective heat transfer within a fluid.

Thus, once convection starts in the molten mantle, the accumulated heat is discharged rapidly. On the ocean floor, the molten mantle, exposed to the surface without the layer of the crust (as was then believed) quickly loses heat. In continental regions, igneous activities are increased as the result of convection within the mantle. When the mantle, having discharged enough heat, solidifies and increases in density again, the continental mass is uplifted isostatically. This, in turn, raises the thick piles of strata, accumulated during the period of depression, into great mountain ranges. Thus, geologic activities at the earth's surface subside into quiescence for some hundred million years till enough heat is again stored for the fusion of the mantle. This is a brief outline of Joly's theory, the so-called "theory of thermal cycles."

What bearing does it have on our main subject, the continental drift theory? If, during the orogenesis, the mantle melted and became a smooth fluid, would not a small force suffice to move the continents? Advocates of the continental drift theory, baffled by the problem of mechanism, welcomed Joly's theory on this score. Joly himself was at first neutral about the continental drift theory; his object in introducing the theory of thermal cycles was to explain the periodicity of geologic revolutions. At the New York Symposium, however, Joly, seeing that his theory could be useful, spoke in favor of the continental drift theory, to the great delight of its supporters.

In Joly's day, data on subcrustal structure and radioactivity were still scarce and Joly actually overestimated the quantity of radioactive heat source in *subcrustal* material. However, his basic idea—that radioactivity within the earth accounts for the periodicity of orogeneses—is valuable, for the radioactive heat within the mantle, small as it is by modern estimate, is still considered today as a possible cause for the periodic occurrence of mountain-building, as we shall now see.

The Theory of Convection Within the Mantle

While we are on the subject of thermal convection, we will look into another theory which is relevant to the mechanism of continental drift: the theory of thermal convection within the mantle. Whereas Joly claimed that thermal convection would be set up in a *melted* mantle, this new theory postulates that convection can occur even in an apparently *solid* mantle. It had been earlier speculated, quite independently of the continental drift theory, that the mantle, although apparently solid, seems to be moving in convective currents which flow at a slow rate on the geologic time scale. This idea, like Joly's theory, was conceived in an attempt to explain the origin of mountains. As we saw in Chapter 1, the traditional theory on mountain formation had been the contraction theory. This declined, however, when closer study revealed that the earth could not possibly have contracted to the degree necessary for the formation of great mountain ranges. Moreover, the discovery of radioactive elements within the earth made it doubtful whether the earth is cooling and contracting at all. As for the periodicity of geologic revolutions, the contraction theory could not even begin to account for that. Thus a fresh start had to be made. Among the various theories offered, one of the most acceptable is the theory of convection within the mantle which we shall now outline according to its chief exponent, D. Griggs (1939).

The basic assumption of this theory (*18*) is similar to Joly's: radioactive heat accumulates within the earth because of the poor conductivity of rocks. This heat causes the substratum of the mantle to expand and become light until, finally, convection starts within the mantle. Once it starts, the heat is effectively conveyed outward. The mantle gradually cools, convection comes to a stop, and geologic activities cease until the next recurrence of convection. This concept—that history repeats itself—is quite similar to Joly's. Look at Figure 2-15 which illustrates this theory. In Stage I, heat is accumulating in the substratum of the mantle, but convection has not yet started. This stage lasts for about 25 million years.

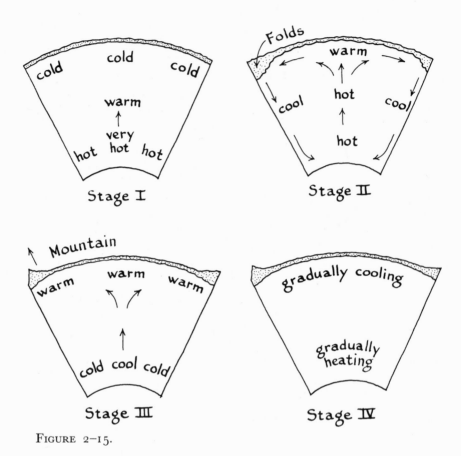

FIGURE 2–15.

Cross-section through the mantle illustrating four stages of Grigg's convection theory. (Adapted from figure 12 in A theory of mountain building,
by D. Griggs, in *American Journal of Science*, *237*, *Sept. 1939*, *p. 636*, with
permission of the author.)

When the substratum is heated to a certain degree, convection
starts within the mantle. Once started, currents develop rapidly
and the heated material rises to the surface. This is Stage II.
During this stage, the currents, in turning downward, tend to drag
and crumple the underside of the crust so that thick piles of folds

are formed (see Figure 2-15, II). The duration of this stage is comparatively short, about 10 million years, at the end of which the currents slow down and come to a stop when the heated material has risen to the surface (Stage III). When the currents stop, the thick pile of geosyncline is no longer held down and will rise isostatically into great mountains. It takes another 25 million years for the current to subside completely. When the orogenesis is over, the next several hundred million years are spent in heating up the cold substratum of the mantle (Stage IV). (It should be noted that although we speak of *currents*, their speed of flow is extremely slow, only 1 ∼ 2 cm per year.) Let us see now how this theory applies to the continental drift theory.

Continents on a Belt Conveyer

A. Holmes of Edinburgh University, a leading figure in modern geology, is noted for his pioneer work in the radioactive determination of geologic periods. He was also the first to suggest that the continental drift might be explained in terms of thermal convection within a "solid" mantle. His idea was put forward at about the same time as the New York Symposium. Look at Figure 2-16. At the early stage of convection, the ascending current rises towards the central part of the primordial continent and forks into two currents going in opposite directions. Such currents tend to pull the continent apart. When the currents are strong, the continent is actually split and begins to drift away with the currents. The advancing front of the current will run into another current coming from the opposite direction and turn downward. In descending, the currents drag down the underside of continental mass which consequently forms a downfold or a fold bulging downward. As soon as the currents stop, this fold rises in response to isostatic force, heaving up a mountain range as it does so. At sea, the descending currents drag down the basalt which forms an oceanic deep. (On the ocean floor, a depression which is more than 7000 meters deep is called an oceanic deep.) The rift between the two split continents becomes an ocean, and the ascending currents, melting into magma,

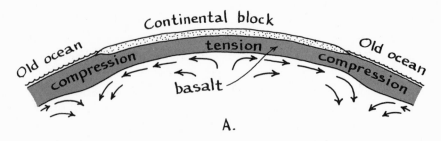

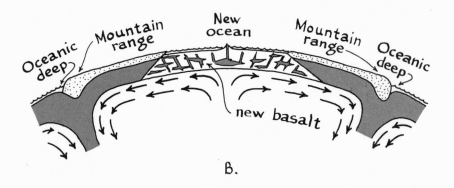

FIGURE 2-16.

Mantle convection as a possible mechanism of continental drift. (Adapted from figure 262 in *Principles of Physical Geology*, by A. Holmes (1945), with permission of Ronald Press Co., New York and Thomas Nelson & Sons, London.)

form new ocean floors and oceanic islands. In this way, the mountains on the advancing front of drifting continents, the oceanic deeps and mid-oceanic ridges (see page 229) are all neatly explained as by-products of the continental drift.

Note that Holmes' theory differs fundamentally from previous theories on the mechanism of continental drift. Until Holmes proposed his theory it was thought that continents *sail* over the mantle and it was difficult to find the adequate mechanism. In Holmes' theory, once we accept the hypothesis that "convection occurs within the mantle," then we may say the continents are *carried*, as it were, on a belt conveyer, the flowing mantle. The engine for this belt conveyer is fed with gravitational energy and thermal energy within the earth.

Of course, if convection does not occur within the mantle, this theory would be a castle built on sand. In this sense, the point of argument has shifted to the possibility of convection within the mantle. Holmes himself emphasizes at the end of his book *Principles of Physical Geology* that:

> . . . purely speculative ideas of this kind, specially invented to match the requirements, can have no scientific value until they acquire support from *independent* evidence.(6)

Support for the Drift Theory Declines

We have seen now how the continental drift theory, introduced by Taylor and Wegener, became the subject of lively discussion among geologists and geophysicists. We have followed the history of this controversy at some length and introduced various new and fantastic theories which emerged in connection with it. Our object was to show what a wide variety of argument was evoked by this single fundamental problem, and how, in the course of controversy, important connections began to emerge among what had seemed to be irrelevant questions.

To review the situation in 1930, the point of argument had shifted considerably from the original theories of Taylor and Wegener, as a result of the appearance of new theories such as Du Toit's theory of a continental ship, Joly's theory of thermal cycles, and Holmes' theory of mantle convection. All these new theories, products of great original minds, were extremely advanced for their time.

Unfortunately, however, data which could provide reasonable and corroborative evidences for these ideas were still lacking. Geophysicists, prone to consider geologists as somewhat less scientific than themselves, and wary of too revolutionary ideas, were not only indifferent but negative in their attitude towards these ideas.

All new theories, good or bad, have one common feature—they differ from the general belief prevailing at the time. It is easy to take a negative attitude and oppose a new idea on the strength of current belief. Most leading geophysicists silently concurred with the orthodox geologists in rejecting the new idea and condemning the continental drift theory as an utter impossibility. And such behavior was considered to be scientific as recently as the 1930's!

H. Jeffreys at Cambridge University is one of the greatest scientists of the present century, in both mathematical physics and geophysics. Jeffreys, however, took a cautious approach toward new theories. In the 4th edition (1959) of his monumental work, *The Earth*, he rejects the continental drift theory with the following words:

> In fact, we can apply to the theory as proposed by Wegener, the words used by Dutton about the thermal contraction theory. "It is quantitatively insufficient and qualitatively inapplicable. It is an explanation which explains nothing which we wish to explain. (*10*)"

Scientists, busy with their own urgent problems, cannot go on wasting time and energy on unorthodox theories. By the end of the 1930's, all that could then be said had been said. The continental drift theory was thus dismissed from people's mind as Wegener's wild dream and was hardly even alluded to in university lectures. The continental drift theory would have died completely, if it had not been for the dramatic emergence, after World War II, of the "independent evidence" emphasized by Holmes.

The nature of progress in science is indeed fascinating. This is not only true for the continental drift theory; the same can be said for the question of thermal convection within the mantle or for radioactivity as the heat source of the earth. As data from observation and measurement increased rapidly after World War II, these ideas, which our predecessors had hit upon with their keen insight, appear again and again in a new guise. Indeed what impresses us is the fact that, in so many cases, our great predecessors had pointed out many decades ago the ideas that are in the limelight today.

We close this chapter on that note then, the apparent death of the continental drift theory. In the next chapter, we will turn to another aspect of earth science, geomagnetism. There we shall see how the "independent evidence," emphasized by Holmes, emerges from an unexpected quarter.

3

The Mystery of the Earth's Magnetism

The Earth as a Magnet

The compass that mariners, hikers, and mountain climbers use to find north, and the horseshoe-shaped magnet that attracts iron, have essentially the same characteristics. Each is a magnet having two ends or poles.

A magnet balanced horizontally in the air always takes a position in which one end points approximately to the earth's geographical North Pole. This end is called the north-seeking or N pole of the magnet; the opposite end is known as the south-seeking or S pole. The like poles of two magnets, N and N or S and S, repel and the unlike poles, N and S, attract each other (Figure 3-1A).

Why does the N pole of a compass needle point to the geographical north? The answer is simple. The earth itself is a huge magnet. Its *S pole as a magnet* is around the *geographical North Pole*. Since unlike poles attract each other, the N pole of a compass needle is attracted towards the S pole of the earth magnet, that is, towards the geographical north. This is illustrated in Figure 3-1B.

Today, the fact that the earth is a huge magnet is common knowledge, taught to school children. But man did not realize this fact until the 16th century. Since then, the study of the earth's

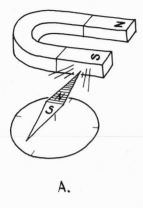

A.

(S. Pole as a magnet)
Geographical North

Geographical South
(N. Pole as a magnet)

B.

FIGURE 3-1.

magnetism has continued steadily, and knowledge and understanding of it have increased greatly. To understand better the significance of magnetism in the consideration of continental drift, we review here some of the basic facts and current beliefs about the earth's magnetism.

A Compass Needle Does Not Point Exactly North

A compass needle tells the geographic north-south direction, but only approximately. The geographic North and South Poles are the two points where the earth's axis of rotation intersects the earth's surface. The compass needle often deviates east or west of the geographic north. This angle of east or west deviation is called *declination*. Figure 3-2 shows the contours of equal declination over the world. Declination is small in Japan, but in the central part of

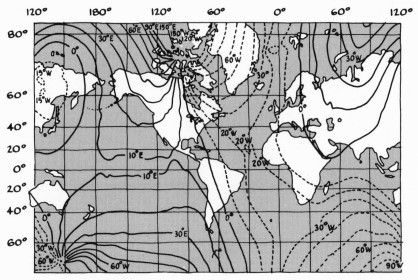

FIGURE 3–2.

Lines of equal declination over the world, 1945. (Adapted from figure on p. 26 in The geomagnetic field, its description and analysis, by E. Vestine, I. Lange, L. Laporte, and W. Scott, *Carnegie Inst. Wash. Pub. 580, 1947,* with permission of the authors.)

Greenland the declination is 60° to the west, so that at that point the compass needle can hardly be said to point north. Only along the contour where the declination is zero degree (see Figure 3-2), does the compass needle point to true north.

A compass needle that is balanced at its center of gravity will not come to rest in a horizontal position. The north-seeking end dips downward in the Northern Hemisphere, as does the south-seeking end in the Southern Hemisphere. (For practical reasons the compass needle is therefore weighted on one side or the other of its center of gravity so that it will rest horizontally.) This angle of deviation from the horizontal is called the *magnetic dip* or *inclination.* The amount of inclination also varies according to location on the earth. This variation is shown in Figure 3-3; there we can see that the inclination is nearly zero at the Equator, but increases with latitude.

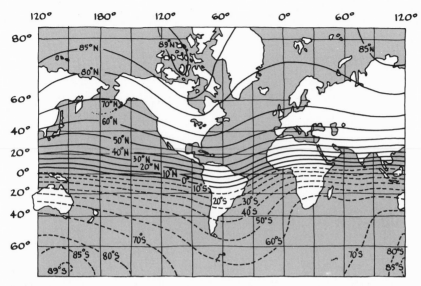

FIGURE 3–3.

Lines of equal inclination over the world, 1945. (Adapted from the figure on p. 27 in The geomagnetic field, its description and analysis, by E. Vestine, I. Lange, L. Laporte, and W. Scott, *Carnegie Inst. Wash. Pub. 580, 1947*, with permission of the authors.)

Beside the declination and the inclination that determine the needle's direction, the earth's magnetic field at any one point is also characterized by its intensity. The intensity of a magnetic field is measured in gauss,* a unit named after the great German mathematician and pioneer of geomagnetic studies, C. F. Gauss. A toy magnet, for instance, can usually produce a magnetic field of several

---

* A gauss is defined as follows. Assume that an electric current of 10 amperes flows in an electric wire placed in a uniform magnetic field the direction of which is perpendicular to the wire. Because of the interaction between the electric current and the magnetic field, the latter exerts a force on the electric wire. The intensity of the magnetic field is defined in terms of this force. When the force is 1 dyne per 1 cm of electric wire, we say that the magnetic field has an intensity of 1 gauss. Actually gauss is a unit of magnetic flux density, but in a vacuum or in air, it gives the intensity of the magnetic field.

tens of gauss. The earth's field is rather weak, about 0.3 gauss near the Equator and 0.7 gauss in polar regions. Figure 3-4 shows the contours of equal intensity over the world.

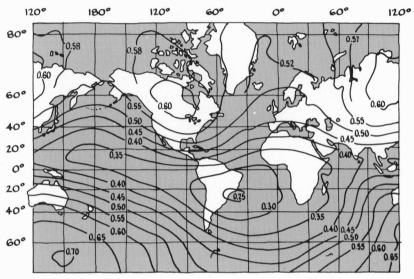

FIGURE 3–4.

Lines of equal total intensity in gauss, 1945. (Adapted from the figure on p. 28 in The geomagnetic field, its description and analysis, by E. Vestine, I. Lange, L. Laporte, and W. Scott, *Carnegie Inst. Wash. Pub. 580, 1947*, with permission of the authors.)

The figures given above indicate *the total intensities* of the geomagnetic field at each locality. To express intensity, its horizontal and vertical components are also commonly used. Look at Figure 3-5. The magnetic field at point A can be represented by an arrow (AC) the direction and the length of which correspond to the direction and the total intensity of the field respectively. Consider a vertical plane containing the arrow AC. On this plane, make a rectangular, ABCD, taking AC as the diagonal and AB and AD in the horizontal and the vertical directions respectively. The arrows AB and AD thus determined are respectively the *horizontal*

and the *vertical components* of the intensity of the magnetic field at
point A.

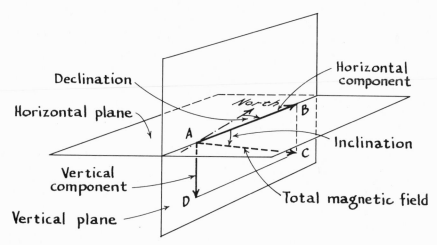

Figure 3–5.

Horizontal and vertical components of the intensity of the geomagnetic
field.

In the 16th century, William Gilbert, physician to Queen
Elizabeth I, cut a piece of magnet in the shape of a sphere and
studied the direction of the magnetic field over its surface. He
found that the distribution of inclination agreed with what was
then known of the earth's field. He concluded from this that the
earth itself was a huge spherical magnet. Later, geomagnetists
found that when the distribution of declination is also taken into
account, the earth's field is best approximated by the field of a
spherical magnet whose axis is inclined at 11° to the earth's geographical axis.

Imagine an infinitesimally small magnet whose poles are in-
finitely close. Such a magnet is called a *dipole* and its field a *dipole
field*. It can be mathematically proved that the field of a spherical
magnet which is uniformly magnetized in one direction is the same

as that of a dipole placed at its center in that direction (see Figure
3-6). The earth's field, then, can be represented by a dipole field
whose axis is inclined at 11° to the earth's geographical axis. The
dipole axis, if extended, intersects the earth's surface at two points

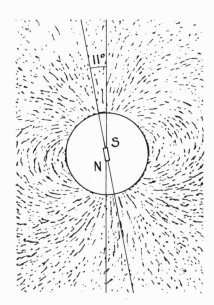

FIGURE 3–6.

The field of a spherical magnet and
a dipole field. Iron filings scattered
around a spherical magnet form
lines of force. The same pattern is
obtained if a dipole is placed at the
center.

situated at $78\frac{1}{2}$°N, 69°W (in northwest Greenland) and $78\frac{1}{2}$°S,
111°E (in Antarctica). These points are called the *geomagnetic poles*
and must be distinguished from the N and S poles of the earth
mentioned on page 94.

Although the dipole field approximates the earth's field, close
study of Figures 3-2, 3-3 and 3-4 shows that the earth's magnetic
field has considerable irregularities so that there are some differences between it and a simple dipole field. This difference is
called the *non-dipole field* or the *geomagnetic anomaly*. Figure 3-7 shows
the anomaly of the vertical component of the intensity of the earth's
field. Note that there are some large-scale anomalies extending over
several thousand kilometers. They are due to causes deep within
the earth (see page 119).

There are also small local anomalies that cannot be shown on a map of this size. These small ones result from local deposits of highly magnetic minerals such as iron ore. Geologists find magnetic "prospecting" a useful technique for locating such ore bodies.

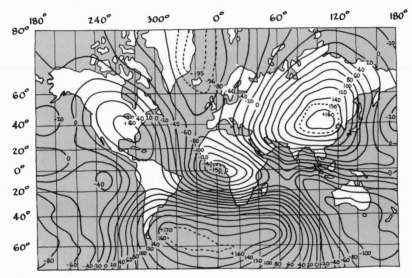

FIGURE 3–7.

Anomaly of the vertical component of the intensity of the geomagnetic field. The unit is 1×10^{-3} gauss. (From The westward drift of the earth's magnetic field, by E. Bullard, C. Freedman, H. Gellman, and J. Nixon, *Roy. Soc. London, Philos. Trans.*, *243*, *67–92*, *1950*, with permission of the authors.)

The Earth's Magnetic Field Changes

Continuous magnetic records of any observatory show that the earth's magnetic field is not constant, but is continually changing. Indeed, there is a slight and fairly regular variation from day to day (called *daily variation*). The variation of declination amounts to a few minutes of arc and the variation of intensity is of the order of 10^{-4} gauss. Figure 3-8 shows the mean daily variation recorded at Kakioka Observatory in Japan.

On some days large disturbances occur, amounting to several degrees in declination and 0.01 or more gauss in intensity. These are called *magnetic storms*. A magnetic storm, which takes days to subside, may be accompanied by disturbances of radio communications and often by the appearance of the aurora in the polar regions.

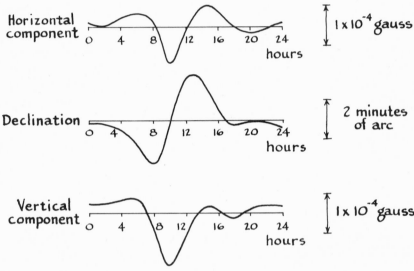

FIGURE 3–8.

Daily variation at Kakioka Observatory, Japan. (From figure 41 in *Chijiki no Nazo*, by T. Rikitake (1957), with permission of the author.)

Since daily variations and magnetic storms are rapid transient changes, it is hardly likely that the cause lies within the earth; in fact, they are caused by electric currents in the upper atmosphere. A few hundred kilometers above the earth's surface, there is a region called ionosphere, where electrons are stripped off from oxygen and nitrogen atoms by radiation from the sun. The positively and the negatively-charged particles (ions and electrons) make the air in the ionosphere a conductor of electricity. Electric currents in the ionosphere produce magnetic fields which cause transient variations in the magnetic field of the earth.

The measurement of slight variations in the earth's field is a delicate matter; for example, electric power lines interfere with the measurement, especially in urban areas where there is more flow of power. Therefore, magnetic observatories must be constructed in remote spots.

As the earth's magnetic field changes constantly, the values of magnetic distribution in Figures 3-2, 3-3 and 3-4 are the averages of those recorded at various stations over a period of time.

The Pattern of the Geomagnetic Field Drifts Westward

Are the recorded average long-range values of the geomagnetic field more or less constant? No, indeed. Figure 3-9 shows the change in declination and inclination at London and at Paris over the past

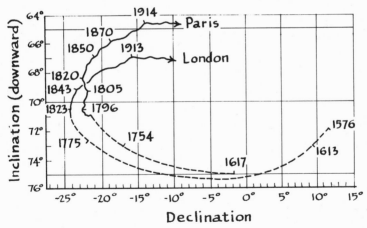

FIGURE 3-9.

Secular variation of the values of declination and inclination in London and Paris. (Adapted from figure 3 by Gaibar-Puertas in *Continental Drift*, edited by S. K. Runcorn (1962), with permission of Academic Press Inc., New York.)

few centuries. Each point on the curve represents the mean values of declination and inclination in the year indicated. This long-term variation is called *secular variation*. Figure 3-10 shows the

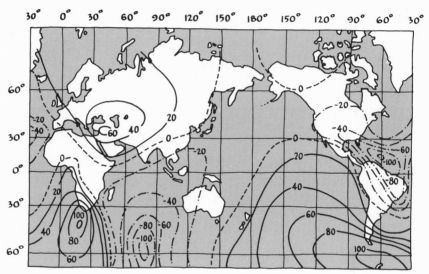

FIGURE 3–10.

Secular variation of the vertical component of the intensity of the geo-magnetic field for the period 1950–1955. The unit is 1×10^{-5} gauss per year. (Adapted from figure 4a in Geomagnetic secular variation during the period from 1950 to 1955, by T. Nagata and T. Rikitake, *J. of Geomagnetism and Geoelectricity*, *IX*, *No. 1*, *March 1957*, *p. 48*, with permission of the authors.)

change in the vertical component of the intensity of the earth's field in one year. While the annual change is very small (of the order of 10^{-4} gauss), the change over several decades may be considerable (see Figure 3-12).

A most significant phenomenon about secular variation is the "westward drift"—that is, the patterns of distribution of the geo-magnetic field as shown in Figures 3-2, 3-4 and 3-7 slowly move westward. For instance, in Figure 3-2, the line of zero declination crosses the Equator at about 65°W, in South America; about four centuries ago, in 1550, it crossed the Equator at about 20°E, in Central Africa, and since then it has moved westward across the Atlantic to its present position.

If we apply the rate of annual variation in Figure 3-10 to the distribution pattern in Figure 3-7, we find that in one decade the pattern shifts slightly west. It is as though we were looking at a revolving light on the shade of which the geomagnetic anomaly has been mapped. The rate of westward drift is about 0.18° in longitude per year; at that rate, the pattern should go around the earth in approximately 2000 years. The speed of westward drift is significant. Unlike magnetic storms which arise from external causes, the large-scale geomagnetic anomalies and their westward drift must be caused within the earth. Generally, changes within the earth occur very slowly, encompassing millions and even billions of years; two thousand years is a very short time in comparison. Indeed, this is an important key to the explanation of the earth's magnetism, as we shall see.

The secular variation is not limited to the declination, the inclination, and the intensity of the earth's field. The *magnetic moment* of the earth's dipole itself changes slightly from year to year. As a magnetic moment is an important concept, we will digress briefly to explain it.

The magnetic moment of a magnet indicates the degree of magnetization of the magnet. To define a magnetic moment, we must specify not only its magnitude but also its direction; the direction is from the S to the N pole; the magnitude is measured in a unit called CGS electromagnetic unit (commonly abbreviated to CGS emu); it is determined as follows.

Imagine a bar magnet 1 cm long. When it is suspended in a magnetic field of 1 gauss, it will align in the field direction (see Figure 3-11A). If we turn the magnet perpendicular to the field direction, as shown in Figure 3-11B, the magnet will tend to turn back to the field direction. In order to keep the magnet perpendicular to the field direction, we must apply some force to its poles. The stronger the magnetization of the magnet, the greater the

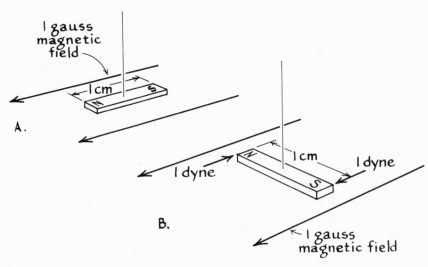

FIGURE 3–11.

Magnetic moment of a magnet. The magnetic moment of the magnet in (B) is 1 CGS emu.

force required. When we need forces of 1 dyne* applied perpendicularly to each pole to keep the magnet from moving (see Figure 3-11B), we say that the magnet has a magnetic moment of 1 CGS electromagnetic unit. Since a magnetic moment has both magnitude and direction, it is represented by an arrow of appropriate length.

The magnetic moment of a magnet can also be estimated from the intensity of the magnetic field that it produces; this is how the magnetic moment of the earth's dipole is determined. At the beginning of the 19th century when Gauss calculated it for the first time, it was 8.5×10^{25} CGS emu. Since then, it has been steadily decreasing (see Figure 3-12); it was measured as 8.0×10^{25} CGS

---

* The unit of force. The force exerted by the earth's gravity on a body having a mass of 1 gram is about 980 dynes.

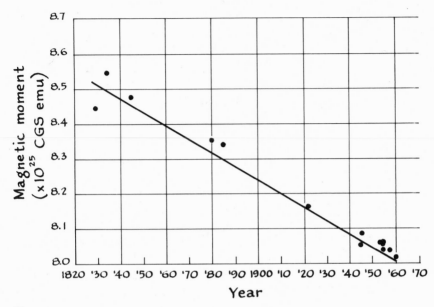

FIGURE 3-12.

Secular variation of the magnetic moment of the earth's dipole from 1829 to 1960 in units of 10^{25} CGS emu. (Adapted from figure 1B in Integral and spherical-harmonic analysis of the geomagnetic field for 1955, part 1, by E. Vestine, W. Sibley, J. Kern, and J. Carlstedt, in *J. of Geomagnetism and Geoelectricity*, XV, No. 2, *47–72, 1963*, with permission of the authors.)

emu in 1960. At this rate, the earth's magnetic field would vanish in 2000 years, and, as a distinguished geophysicist remarked, "all geomagnetists would find themselves out of work in the near future." Meanwhile, we do not know whether the magnetic moment of the earth's dipole will continue to decrease, or why it does so at all. We do know that the earth's field is changing.

Why Is the Earth a Magnet?

At the beginning of this chapter, we asked why the compass needle points to the north, and the answer was that the earth itself is a magnet. But isn't that like saying that it rains because the weather is foul? We ought to have explained *why* the earth is a magnet. Why? The answer is elusive. Geophysics still finds this question one of the greatest mysteries of science. To understand the various hypotheses offered as a solution, we need to know some basic facts about the magnetic property of a substance.

Can Any Substance Be a Magnet?

Some substances, like iron and nickel, are capable of high magnetization; others, like copper and aluminum, are not. Substances are grouped into three categories according to their magnetic properties, the ferromagnetic, the paramagnetic, and the diamagnetic.

Iron, nickel, and cobalt are easily magnetized. When brought near a magnet, the field of that magnet induces magnetism in them. The end of the metal nearest the N pole of the magnet becomes the S pole, as in Figure 3-13. Such highly magnetic substances are

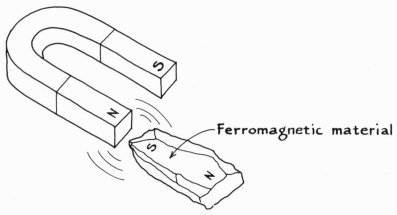

Ferromagnetic material

FIGURE 3-13.

called ferromagnetic materials. They include non-metallic substances such as ferrite, which is used as a magnet for holding paper against a steel blackboard. Non-metallic ferromagnetic substances are important in rock magnetism, as we shall see in the next chapter.

However, most substances around us are non-ferromagnetic; that is, they are paramagnetic or diamagnetic. A paramagnetic substance, when brought into a magnetic field, is magnetized in the direction of the field, but only weakly. It is attracted by a magnet, but only a most sensitive instrument can detect it. Paramagnetic substances include platinum, aluminum, and oxygen.

Diamagnetic substances are perverse in that when they are brought into a magnetic field they magnetize in the opposite direction, repelling instead of being attracted to a magnet. Diamagnetic substances, which are numerous, include gold, silver, copper, lead, carbon dioxide, and water. The intensity of magnetization in a diamagnetic substance is comparable to or weaker than that of a paramagnetic substance.

Why is it that some substances are ferromagnetic and others are not? This question brings us to the subject of the behavior of atoms and electrons in a substance.

What Causes Ferromagnetism?

The magnetic property of a substance can ultimately be attributed to the motion of electrons in the atoms that make it up. You will recall from Chapter 2 that an atom is composed of a nucleus surrounded by one or more rapidly-moving electrons. The motion of an electron is two-fold. It revolves around the nucleus; this is the orbital motion. It also rotates on its axis; this is the spin motion.

The orbital and the spin motion of the electron each produces a magnetic field. This is the basic cause of magnetism; an electric current produces a magnetic field because the former is in fact just a flow of electrons.

As the electron in orbital motion produces a magnetic field, it is convenient to consider it as a kind of magnet, and to express its magnetic property in terms of the magnetic moment of the orbital motion. Similarly, we speak of the magnetic moment of the spin motion. The magnetic moment of an atom is the combined effect

of the magnetic moments of the orbital and the spin motions of all
the electrons that it contains.

Electrons are arranged around the nucleus in successive subshells. Look at Figure 3-14 which shows schematically the distribu-

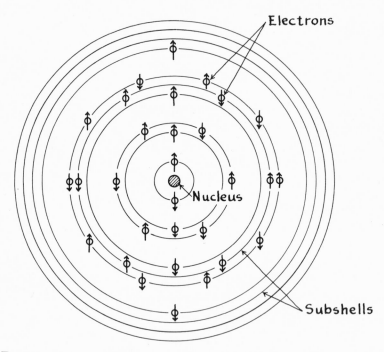

FIGURE 3–14.

Schematic drawing of the distribution of electrons in the atom of zinc.
The arrows show schematically the directions of the spin magnetic mo-
ments. (Adapted from figure 10a in *Magnetism, From Lodestone to Polar
Wandering*, by D. S. Parasnis (1961), with permission of Harper & Row,
New York.)

tion of electrons in the atom of zinc. Each subshell can accommodate only a limited number of electrons, and the number of
electrons that a subshell may accommodate when it is filled is
always even; for instance, the innermost subshell can take only two

electrons, the next one two, the third one six and so on (see Figure 3-14). The atom of hydrogen has only one electron, so its innermost subshell is *incompletely* filled; in the atom of helium which has two electrons, the subshell is *completely* filled.

In atoms with completely full subshells (such as helium or zinc), half the electrons spin in one direction and the other half in the opposite (see Figure 3-14). Therefore, the spin magnetic moments of electrons cancel each other completely. Moreover, these atoms are characterized by an inward symmetry which may be visualized by imagining half the electrons to revolve in one direction and the other half in the opposite. Therefore, the orbital magnetic moments also cancel one another completely. Hence, the atom has no net magnetic moment.

When a substance composed of such atoms is brought into a magnetic field, the direction of the orbital magnetic moment is slightly changed; it tilts, so to speak, in the direction opposite to that of the applied field. Because of this change, the orbital magnetic moments of electrons no longer cancel one another completely, and the substance acquires a magnetic moment in the direction *opposite* to that of the applied field; such a substance would repel a magnet. This is diamagnetism.

In other substances, the electronic subshells are *in*completely filled, so that the orbital and the spin magnetic moments do not cancel each other completely; the atom has a net magnetic moment. Normally, however, an atom has a thermal energy and is constantly vibrating and rotating in a random manner. Because of the motion, the magnetic moments of atoms are oriented quite at random as shown in Figure 3-15A. Therefore the substance as a whole has no net magnetic moment in any particular direction. When such a substance is brought into a magnetic field, the magnetic moments of atoms tend to align in the field direction. The tendency, however, is opposed by the thermal motion of atoms and the alignment is not complete; hence the substance acquires only a weak magnetic moment in the field direction. This is paramagnetism.

Ferromagnetism is possible only when the atoms influence each other, forcing the magnetic moments of neighboring atoms to parallel their own. (See Figure 3-15B.) This is called *exchange interaction*. When the exchange interaction is stronger than the random thermal motion of atoms, the substance is ferromagnetic. While a ferromagnetic substance is *un*magnetized, the alignment of magnetic moments is not uniform throughout the substance; that is, the substance can be divided into blocks called domains

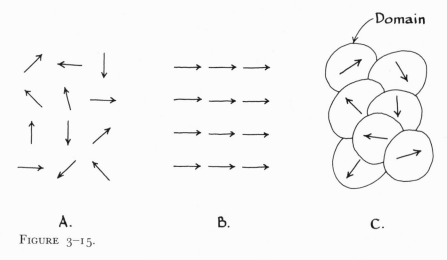

A.   B.   C.

FIGURE 3–15.

Schematic drawing of the directions of magnetic moments in paramagnetic and ferromagnetic substances.

and within each domain, the magnetic moments of atoms are aligned in one direction, but the direction of alignment differs from one domain to another. (See Figure 3-15C.) Hence, two pieces of unmagnetized iron do not attract each other. Once placed in a magnetic field, the domains set themselves parallel to the field direction and the substance as a whole acquires a strong magnetic moment. The domain theory, proposed by a French scientist, P. Weiss, early in the present century, has been confirmed by many experiments.

Ferromagnetic substances differ from the paramagnetic and the diamagnetic not only in the intensity of magnetization, but also in the lasting effects. With the removal of the magnetic field, paramagnetic and diamagnetic substances lose magnetization completely, but a ferromagnetic substance retains magnetization to a considerable degree; it has become a *permanent magnet*. Magnetization that remains in substances after the removal of the magnetic field is called, sensibly, a *remanent** magnetization*. The horseshoe-shaped magnet, for instance, is a piece of ferromagnetic substance that has been magnetized by the application of an extremely strong magnetic field, and has retained a large remanent magnetization.

As we noted above, a substance is ferromagnetic only when the exchange interaction is stronger than the thermal energy of atoms. The latter varies with temperature. The thermal energy of an atom is zero at $-273°C$ (this temperature is called absolute zero) and increases with temperature until at a particular temperature it exceeds the exchange energy; the magnetic moments of atoms take random directions, and the substance ceases to be ferromagnetic. This temperature is called the Curie point, named for Pierre Curie who did pioneer work in the study of magnetic materials at the end of the 19th century.

With this preliminary knowledge, let us now examine various hypotheses on the cause of the earth's magnetism.

The Permanent-Magnet Hypothesis

There are two ways in which a magnetic field can be produced, by a magnetized substance (a magnet) or by electric currents. Of the two possibilities, the former is simpler to imagine for anyone who has seen a magnetized piece of matter and its effects. Indeed, our ancestors thought that the earth's magnetic field was caused by a great magnet buried in the earth. This is known as the permanent-magnet hypothesis. After all, the earth contains iron-ore deposits, and, what is more, the earth's core (see page 38) is

---

* A technical term which might be considered a variation of "remnant."

said to be composed of iron and nickel, which are ferromagnetic. If this core, with a radius exceeding 3400 kilometers, is actually a permanent magnet, the earth's main field can be easily accounted for. Furthermore, the geomagnetic anomalies can be explained if we assume a certain lack of homogeneity within the core.

This explanation, however, has a fatal defect. Ferromagnetic substances, such as iron and nickel, lose their ferromagnetism above the Curie point. The Curie point is 770°C for iron and 360°C for nickel. If the temperature in the core is higher than these figures, iron and nickel cannot retain their ferromagnetism. We will deal with the temperature of the earth's interior in Chapter 6; it is sufficient to say here that, the outer core is almost certainly in a liquid state, as we explained in Chapter 1. If iron and nickel in the outer core are in a liquid state, the temperature in the outer core must be fairly high, for we know from experiments that at the earth's surface, the melting points for iron and nickel are 1535°C and 1453°C respectively. In general, the melting point of a substance increases with increasing pressure. In the core, where the pressure exceeds 10^6 bars,* the melting points for iron and nickel must be considerably higher than at the surface. Therefore, since it is quite certain that the core is hotter than 2000°C, iron and nickel could not possibly retain their ferromagnetism. Therefore, the core of the earth cannot be a permanent magnet.

It was once suggested that the permanent-magnet hypothesis might hold because the Curie point may rise under pressure and the pressures at the core are tremendous. Experiments, however, showed that though the Curie point does indeed rise with increasing pressure, it does so only at the rate of 1°C per 10^5 bars. Thus, the Curie point for iron at the surface of the core would be no higher than 780°C. This would not save the hypothesis at all. Also, the permanent-magnet hypothesis can be set aside more simply, for, if the outer core is really in a liquid state, no liquid substance

---

* 1 kilobar = 1000 bars. 1 bar = 10^6 dynes per cm². 1 bar is almost equal to 1 atmospheric pressure.

is known to be magnetic, nor, theoretically, is such a substance likely to exist.

Is it likely that the mantle and the crust, instead of the core, are permanently magnetized? In the case of the mantle, the high temperature would present the same problem, though in the crust, where the temperature is generally not above several hundred degrees centigrade, it would not. As late as the 1950's there were attempts to attribute the earth's magnetic field to the magnetization of crustal rocks by external fields such as magnetic storms. This theory was overthrown when it was demonstrated that the magnetization of the crust is too weak to have produced the earth's field.

Other Hypotheses on the Cause of the Earth's Magnetism

If there are electrically charged particles at the surface of the earth or within it, and rotating with it, they would constitute a kind of electric current and hence would produce a magnetic field. Some scientists claimed that this caused the earth's magnetism. However, to produce the observed geomagnetic field, electric currents of the order of 10^9 amperes are required. This would imply a tremendously strong electric field at the earth's surface; such a field does not exist.

Then there was the attempt to explain the earth's field as the result of the gyromagnetic effect. When a substance is rotated, the orbital and the spin motions of its electrons would be affected; this would change the magnetic moment of the substance; if it had initially been zero, the substance would acquire a magnetic moment as the result of rotation.

This theoretical speculation was confirmed by S. J. Barnett in experiment in 1914, and is thus sometimes called *Barnett effect*. Barnett rotated a ferromagnetic substance at a high speed; although the substance did indeed acquire a magnetic moment, it was very weak and difficult to measure. The earth is mostly composed of non-ferromagnetic substances, and its rotation, one turn per day, is relatively slow. Hence we can hardly expect a strong gyromagnetic effect; in fact, the magnetic field produced by the effect would be ten billion times weaker than the actual field. The gyromagnetic effect cannot be the cause of the earth's magnetism.

A Negative Experiment

One of the most dramatic attempts to explain the geomagnetic field by the rotation of the earth was made by an English Nobel laureate, P. M. S. Blackett in 1947. His great stature as a scientist

FIGURE 3–16.

P. M. S. Blackett.

attracted much attention to his hypothesis. The sun and some stars (for instance, 78 Virginis, the 78th star of the constellation *Maiden*) are known to possess magnetic fields. Thus Blackett theorized that any massive celestial body acquires magnetization from rotation, regardless of whether it is composed of ferromagnetic materials. He suggested that this was a fundamental, though unknown, principle of physics. Further, he claimed that the observed intensity of the earth's field, as well as that of the solar field (then believed to be 50 ∼ 60 gauss), and that of a certain star (8000 gauss) can each be derived theoretically from the mass and the speed of rotation of the respective body.

Sometime later, however, accurate measurements of the solar field revealed that it not only varies with time, but also that it is weak in intensity (several gauss). Blackett's theory, then, would not apply to the solar field.

If, as Blackett claimed, rotation of a celestial body directly causes its magnetization, the earth in its total bulk must be responsible for its own field. Then the intensity of the magnetic field must *decrease* with depth. This was tested by S. K. Runcorn (1950–1951) in English coal mines and by a Danish research vessel, *Galathea* (1950–1952), in the Pacific Ocean; both results were unfavorable to Blackett's theory.

While other scientists were testing the validity of his theory, Blackett was busily engaged in an elaborate experiment to prove his own theory. He placed a gold cylinder in his laboratory which naturally rotated with the earth. The magnetic field produced by the earth is less than 1 gauss in intensity; how was he to measure the field that might be produced by a small mass of gold in a laboratory? Blackett spent years refining an instrument called an "astatic magnetometer" (see Figure 3-17) and he succeeded in raising its sensitivity almost to its theoretical limit where it could measure one ten-millionth of a gauss. When he had at last performed his experiment, he himself, to his own credit and disappointment, published the fact that his theory was incorrect; the expected magnetic field had not been produced after all. He gave a full account of the experiment in an article "Negative Experiment" published in 1952.

Blackett's work had remarkable results nevertheless. Later, he made excellent use of his astatic magnetometer in the study of rock magnetism and, through it, played a brilliant part in the revival of the continental drift theory (see Chapter 5). Nor can we overlook the contribution that he made in publishing a precise

FIGURE 3–17.

Astatic magnetometer.

account of his magnetometer, which enabled other geoscientists to construct the highly sensitive instrument they needed for further study of rock magnetism.

The Earth as a Dynamo

Thus, every attempt to prove that the earth is a permanent magnet failed. There remained the possibility that the geomagnetic field is caused by electric currents within the earth.

The amount of electricity necessary to account for the geomagnetic field is of the order of 10^9 amperes. As stated in Chapter 1, the crust is composed of granite and basalt, and the mantle probably of peridotite or eclogite. All these are poor electric conductors. It is just conceivable that their conductivities are raised under the high temperatures of the earth's interior, but even so, they would be low compared with the conductivity of metal. The core of iron and nickel, on the other hand, is a good conductor. Therefore, if the geomagnetic field is really caused by electric currents in the earth, these currents must flow in the core.

The speed of secular variation supports the core theory. As noted earlier (see page 106), the magnetic moment of the earth's dipole has decreased 5% in the past hundred years, and the pattern of the non-dipole field drifts westward by about 0.18° annually. Thus, whatever causes the geomagnetic field is subject to considerable changes in the relatively short periods of decades and centuries. Such rapid changes could only occur in a liquid core, for, naturally, large-scale changes can take place more rapidly in a mobile liquid than in a solid.

The question then boils down to this: how can electric currents of the order of 10^9 amperes be generated and maintained in the core?

Imagine a dynamo in the earth's core. Electricity for our daily use is generated by a dynamo in a power plant. A dynamo has a fairly complex structure, composed of coils, brushes, and a rotor rotated at high speeds by water or steam power. The earth's core is merely a mass of fluid iron with none of these devices. Is it possible that the fluid iron acts as a dynamo? This sounds plausible in the abstract, but the question of its concrete structure raises problems.

Some scientists have suggested that the electricity in the core is supplied by thermoelectric currents. Figure 3-18 illustrates an example of a thermoelectric current. A copper wire and a constantan wire (constantan is an alloy of copper and nickel) are joined at two ends to form a ring. When one end is heated, an electric current flows through the ring in the direction of the arrow. This is the thermoelectric current. It arises when the junctions of two different conducting materials are kept at different temperatures. A similar process may occur in the core. If there is a thermal convection in the core, there would be some differences of temperature in it. As the mantle and the core are made of different substances, it is conceivable that thermoelectric currents originate at their boundary. It is not known, however, whether the current is quantitatively sufficient to act as a dynamo. The hypothesis of the thermoelectric current as the cause of the earth's magnetism has not been worked out in detail; it has not been

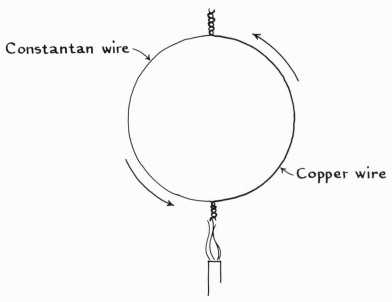

FIGURE 3-18.

Thermoelectric current. The arrows indicate the direction of the current.

disproved but neither has it been accepted. A theory that *has* been worked out in considerable detail and looks more promising than any other is the dynamo theory to which we will turn now.

The Mysterious Dynamo: a Clue

The geomagnetic field is subject to secular variation within short periods. However, in the long run, say over a million years, the earth seems to maintain a more or less stable dipole field (see Chapter 4, page 142). Thus, the core-dynamo theory to be accept-able must account for long-term stability of the dipole field as well as for secular variation. It is not sufficient, therefore, to assert that, from some accidental cause, electric currents of the order of 10^9 amperes are now flowing within the core. The core, although a good electric conductor, has some electric resistance. If we stir water in a tub, making a current and then letting it alone, the current will stop gradually. Similarly, if the electric currents in the core were left alone, the initial electric energy would change to thermal energy because of the electric resistance of the core, and the currents would vanish in a few thousand years. To maintain the currents, the loss due to resistance must be made up by new electric currents. Since there is no external source of supply, the electric currents in the core must be self-supplied; the core must generate electricity on its own. This would make it a wonderful dynamo, indeed.

Suppose electric currents are flowing within the core and that a dipole field exists in consequence. As we shall soon see, we can assume, on the other hand, that there would be a fluid motion in the core from thermal convection. It is a principle of electromagnetism that fluid metal with high electric conductivity in motion within a magnetic field induces a new electric current. A new electric current would produce a new magnetic field. Now suppose this new magnetic field *reinforces* the initial field. Then does it not follow, once we assume the existence of the initial electric currents which produce the dipole field, that the core could, through its own mechanism, maintain this dipole field permanently, supplying new currents and hence the reinforcing fields? If so, the core would be a *self-exciting dynamo*. A highly excitable person, once excited by some stimulus, gets more and more excited at his own excitement and goes off into an apparently endless self-intoxication. In our jargon, that person would be a *self-exciting dynamo*.

The idea that the core is a self-exciting dynamo—called the dynamo theory—was first suggested in 1919 and has steadily gained plausibility.

Ancient men dreamed of the so-called "permanent engine," a perpetual motion machine whose operation is forever maintained through its own mechanism. With such an engine, automobiles and airplanes would work forever without gasoline. Men through the centuries tried to invent perpetual motion machines, but in the 19th century, establishment of the concept of energy showed the idea to be intrinsically impossible. Today, no sane—or informed—

person dreams of a perpetual motion machine. The self-exciting dynamo so far may sound as wild-eyed as the perpetual motion machine, but actually it is not. In the self-exciting dynamo of the core, energy is constantly consumed and a new supply is ever drawn from the energy of fluid motion. The question of the source of this energy will be taken up at the end of the chapter.

The Disc Dynamo

The mechanism of the self-exciting dynamo is fairly complex. Let us start with a simple example, the disc dynamo devised in 1955 by E. C. Bullard of Cambridge University, one of the leading proponents of the dynamo theory. Look at Figure 3-19. In a disc

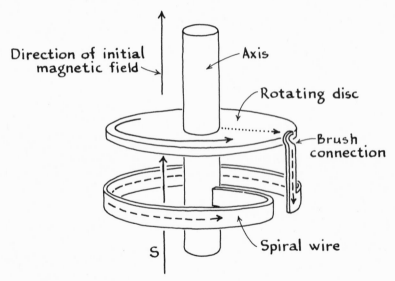

FIGURE 3–19.

A disc dynamo.

dynamo a metallic disc rotates around a metallic axis. Below the disc is a spiral electric wire, one end of which is attached to the axis and the other end to the rim of the rotating disc by means of a brush. Suppose we are given a magnetic field in the direction indicated by the arrow, and that the disc is rotating in the direction of the arrow. When a conductor (the disc) rotates in a magnetic field, an electric current is induced within the disc in the direction indicated by the dotted line. The current flows through the brush to the spiral wire in the direction of the broken line. The current produces a magnetic field in the direction of the arrow S, thus reinforcing the initial field. The disc dynamo is self-exciting. As long as the disc is rotated at a sufficiently high speed, the initial magnetic field is maintained without decaying. Such is the mechanism of the self-exciting dynamo.

It is clear at a glance that the disc dynamo is not a permanent engine: the disc must be rotated to keep the dynamo working. The rotation of the disc corresponds to the fluid motion in the core; the earth dynamo therefore works at the expense of the energy of fluid motion.

It is evident that a self-exciting dynamo requires an initial magnetic field. Without it, the rotation of the disc produces nothing. The same can be said of the earth dynamo. Once the initial field, however weak, is given, it will be reinforced, and becomes a stronger field. What is the initial field in the case of the earth? According to the proponents of the dynamo theory, it may be the extremely weak magnetic field which is said to exist everywhere in the universe; or perhaps it springs from the thermoelectric currents in the core. Note that, while the thermoelectric currents may be too weak to act as the dynamo itself (see page 119), they could be great enough to provide the initial magnetic field.

Dynamo in the Core

The dynamo theory was originally suggested by J. Larmor of England in 1919. After World War II, W. M. Elsasser and Bullard followed up this suggestion and worked out the theory in considerable detail (15). They have shown, in a fairly concrete manner, that in the core of the earth, fluid motions act like the rotating disc in the disc dynamo. The main motion of the core is

FIGURE 3–20.

W. M. Elsasser.

a rotation in accordance with the earth's rotation around the earth's axis. In addition, there are secondary motions in the fluid core. The fluid core of iron and nickel must contain radioactive elements which generate heat, and a sufficient quantity of heat would initiate thermal convection (see page 85). This would produce a current which ascends from the deeper part of the core to its surface and then descends again (see Figure 3-22A). For simplicity, let us call it the *convective current*.

Consider the convective current descending within the fluid core which is itself rotating. When a rotating mass moves towards the axis of rotation, the speed of rotation increases. That is why a skater who is pirouetting with her arms stretched out, folds them when she wants to pirouette faster. In short, the inner part of the liquid core rotates more rapidly than the outer part. Therefore, when viewed from the outer part, the inner part rotates eastward

FIGURE 3–21.

E. C. Bullard.

as shown in Figure 3-22B. This differential rotation of the liquid in the core constitutes the *parallel current.*

How is it that, given an initial dipole field and these two currents, the core becomes a self-exciting dynamo in which the initial field is continually reinforced? The process is complex and may be hard to grasp intuitively, even with the aid of illustrations. Still, we will attempt an explanation.

First, it will help to remember the following two basic laws of magnetism. Look at Figure 3-23. Suppose an electric current flows in the direction of the dotted arrow. It produces a magnetic field in the direction of the solid arrow. The direction of the magnetic field is always that in which a corkscrew advances when the screw is turned along the direction of the electric current. This will be evident in the working of the core dynamo as we shall see.

Figure 3-24 illustrates the second law. Let the solid arrow represent the direction of a magnetic field. Suppose, within the field, an electrical conductor moves in the direction of the broken arrow. Then an electric current flows in the conductor in the direction

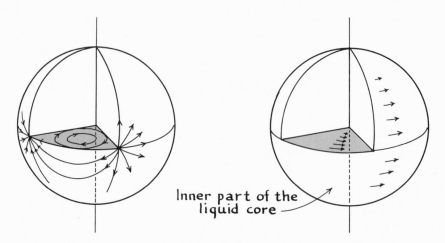

A. Convective current   **B. Parallel current**

FIGURE 3–22.

Two types of fluid motion in the core.

of the dotted line. This is the direction in which a corkscrew advances when the screw is turned so as to bring the direction of motion into coincidence with the field direction. This law, when applied to the core dynamo, presents a somewhat more complex picture than the present example.

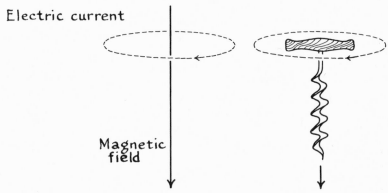

FIGURE 3-23.

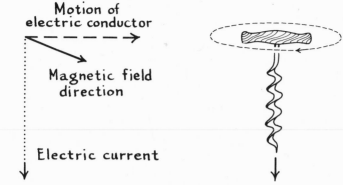

FIGURE 3-24.

Look at Figure 3-25A. Here are the initial dipole field and the parallel currents. The electric currents induced thereby flow in the direction shown by dotted lines on the cross-section. They take the form of two doughnuts, one in the Northern and the other in the Southern Hemisphere. Within the doughnuts, magnetic fields are produced; their directions, shown by the two broken arrows in Figure 3-25B, differ in the Northern and Southern Hemispheres. In both hemispheres, however, the direction of the field is parallel to the surface of the core and confined within its boundary so that, however intense the field may be, its existence is not apparent to those at the earth's surface. Bullard estimated the intensity of this field at 400 gauss, which is nearly one thousand times greater than the intensity of the field at the earth's surface.

Now for the convective currents (see again Figure 3-25B). When convective currents flow in these magnetic fields, as shown by solid arrows in Figure 3-25B, they induce electric currents shown by the dotted arrow. These currents produce the queerly shaped magnetic fields shown by broken arrows in Figure 3-25C. These fields and the parallel currents produce another electric current which, in turn, produces the magnetic fields shown in Figure 3-25D. Finally, the combination of these fields and the initial convective currents produces the electric currents shown by the dotted line in Figure 3-25D which, lo and behold, produces a magnetic field

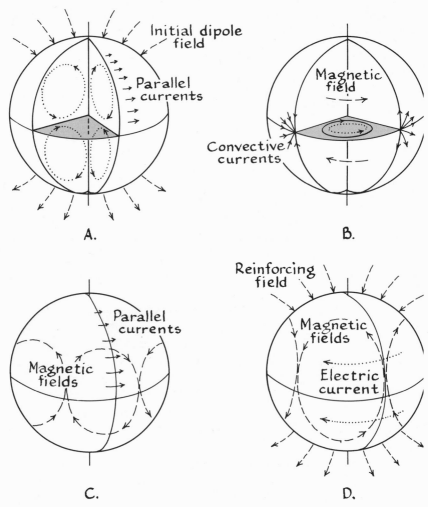

FIGURE 3-25.

Mechanism of the core dynamo according to Bullard. Only the core of the earth is shown in each figure.

which reinforces the initial dipole field. This is an extremely complex and even roundabout procedure but, according to the principles of electromagnetism, this sort of thing *can happen.*

H. Takeuchi and Y. Shimazu (1952) and Bullard (1954) showed that the core-dynamo can work through this mechanism if we assume certain, and not unreasonable, values for the electric conductivity of the core and the speed of fluid motion (*2, 15*). The speed of fluid motion need be no faster than 10^{-2} or 10^{-1} mm/sec.

The reader doubtless has already noticed that Bullard has made several assumptions convenient for his theory. He ignores the possibility that the combination of a certain field and a certain current may bring about results other than the desired currents and fields. He also ignores the electric currents and magnetic fields which may be produced by undesirable elements acting in concert. What would happen if all these possibilities were taken into consideration is not yet known for certain. Therefore, many able scientists still distrust the dynamo theory. They suspect that the Elsasser-Bullard core-dynamo seems possible only because its authors are shutting their eyes to unfavorable factors. The core is after all uniformly composed of metal without electric wires, coil, or brush. Is it possible for such a substance to act as a self-exciting dynamo?

In 1958, G. Backus and A. Herzenberg, working independently, each showed that it was possible to postulate a pattern of motions in a conducting fluid in such a way that it acts as a self-exciting dynamo (*15*). In each case, the motions were physically improbable; however, rigorous mathematical solutions were obtained as was not the case with Bullard's solution. This was a step forward in the development of the dynamo theory.

In 1963, F. J. Lowes and I. Wilkinson succeeded in constructing a self-exciting dynamo without coil or brush (*15*). Their dynamo is shown schematically in Figure 3-26. Two cylinders, made of an iron alloy, rotate in cylindrical cavities in a stationary block of the same material. The cylinders are in electrical contact with the surrounding block through a thin layer of mercury. The axes of the two cylinders are at right angles.

An initial magnetic field is given in the direction of arrow H_I,

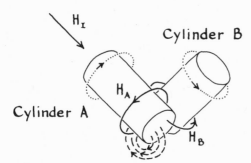

FIGURE 3–26.

Dynamo of Lowes and Wilkinson. (After figure 3 in Geomagnetic dynamo: a laboratory model, by F. J. Lowes and I. Wilkinson, in *Nature*, *198*, *1963*, *p. 1159*, with permission of the authors.)

that is, along the axis of cylinder A. When the cylinder is rotated in the direction of the dotted arrow, an electric current is induced in the direction of the broken arrow. The current produces a magnetic field in the direction of arrow H_A which, as the figure shows, is directed along the axis of cylinder B. Because the axes of the two cylinders are at right angles, the magnetic field produced by one cylinder is always directed along the axis of the other. Therefore, the field produced by cylinder B (shown by arrow H_B) reinforces the initial field. If the cylinders are rotated at a sufficiently high speed, the induced field is as large or larger than the initial field so that the latter is no longer needed; the dynamo is self-exciting.

Figure 3-27 shows the result of an experiment with this dynamo. One cylinder was rotated at the constant speed of 2650 turns per minute and the speed of the other was gradually increased from zero. At the speed of about 1500 turns per minute, there was a sudden increase in the intensity of the induced field outside the block, indicating the onset of dynamo action. The experiment confirms what Backus and Herzenberg had proved mathematically, that a dynamo can be self-exciting without coil or brush.

Finally, a word about the energy source of the core dynamo. We emphasize again that the core dynamo, although self-exciting, is not a perpetual motion machine. The loss of energy from electric resistance must be made up by the kinetic energy of fluid motion which induces new electric currents in the presence of a magnetic field. Where does the kinetic energy (the energy of motion) come from? It comes from the thermal energy and the

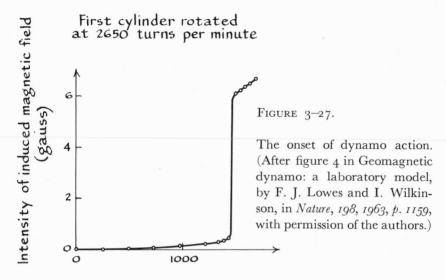

First cylinder rotated
at 2650 turns per minute

Intensity of induced magnetic field
(gauss)

Speed of rotation of second cylinder
(turns per minute)

FIGURE 3–27.

The onset of dynamo action. (After figure 4 in Geomagnetic dynamo: a laboratory model, by F. J. Lowes and I. Wilkinson, in *Nature, 198, 1963, p. 1159,* with permission of the authors.)

energy of the earth's rotation, for as we noted earlier, convective currents are caused by thermal convection, and the parallel currents by the combination of the convective currents and the earth's rotation.

The energy consumed in the core-dynamo is estimated at 2×10^{10} cal/sec; the thermal energy required to start a thermal convection is one order of magnitude greater than that. To generate that much heat, the radioactive elements in the core need be about 1/1000th of those contained in the crust. Bullard thought it possible for the core to contain that much radioactive heat.

Since then, precise measurements of meteoritic iron has revealed that its radioactive content is extremely small—about 1/100,000th of that in the earth's crust. Meteoritic iron and core iron may not be identical, but one might expect them to be roughly comparable (see Chapter 6). If we accept the comparison, we must conclude that the radioactivity of the core cannot be the energy source of thermal convection and hence of the core-dynamo.

Recently, an additional thought has been brought to bear on the question. There are reasons to believe that while the outer part of the core is liquid, the inner part is solid (see page 38), and that the solid part is slowly growing. If that is so, as part of the liquid core goes through the transition to a solid state, thermal energy would be released. J. Verhoogen has suggested that this thermal energy could be great enough to set in motion the thermal convection currents.

Nevertheless, the question of the primary energy sources for the core-dynamo remains an intriguing and important one.

A Summary

What is the origin of the geomagnetic field? Many a hypothesis has been offered and rejected. The fairest answer at present would be that the geomagnetic field is caused by electric currents within the core, and that these currents are probably induced and maintained by a mechanism such as the self-exciting dynamo. The dynamo theory is certainly by far the best yet offered. It is not perfect, but it surpasses all the others by a wide margin.

This chapter has mainly dealt with some basic facts about the *present* geomagnetic field and the pros and cons of various hypotheses on its origin. In the next chapter, we shall see how the *past* geomagnetic field has been retained in rocks and ancient pottery. Such "fossil" evidence of the geomagnetic field sheds a new light on the ancient positions of continents and how they may have drifted in the geologic past.

4

Fossil Magnetism

Ancient Fires and Modern Science

There is an anthropologist at the University of Tokyo, N. Watanabe, who travels around the country looking for sites of ancient fireplaces where primitive men baked their earthenware pottery and prepared their meals. When he finds such a place he first measures with compass and level the exact orientation of any shards of the pottery and any baked lumps of the earth that seem to him suitable for his purpose. Then he takes the fragments back to his laboratory, where the direction of the magnetization of each lump or shard is measured with a magnetometer.

Through this process, the anthropologist expects to determine the age of the fireplace. The earthenwares and the lumps of earth baked by the fires generally contain small amounts of ferromagnetic substances that, upon cooling, must have been magnetized in the direction of the geomagnetic field that existed at that time. If we know how the direction of the geomagnetic field—declination and inclination—has changed in the past several thousand years, then we can infer the age of the fireplace from the direction of its magnetization.

This is not as easy as it sounds; as the measurement of secular variation was started only three centuries ago, we do not know the history of the geomagnetic field prior to that. Therefore, simply knowing the magnetic direction of baked earth may not suffice to determine its age. First, the history of earlier secular variation must be determined. This can be done by measuring the magnetic direction of volcanic rocks, baked earth and earthenware whose dates have already been determined from historical documents or by radioactivity. Such work is under way in England, France, Japan, the United States and elsewhere. Archaeological sites in the Mediterranean, America and elsewhere have yielded extensive data, thanks to the cooperation of archaeologists with geophysicists. An estimate of the change of the direction of the geomagnetic field in Japan is given in Figure 4-1. The change has been inferred mainly from the magnetic direction of lava flows whose dates of eruption are known. Both declination and inclination change with

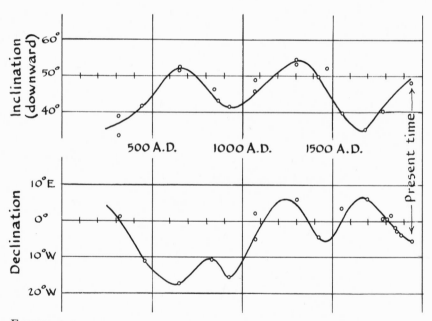

FIGURE 4–1.

Estimated change of declination and inclination in Japan. (Adapted from T. Yukutake and N. Watanabe.)

time with some such irregular, wave-like motion, as Figure 4-1 shows. Thus, the history of secular variation is slowly being worked out.

In all this, we assume that the direction of the past field is fixed and retained in rocks and baked earth—that those are, in effect, fossils. A fossil of the geomagnetic field is the remanent magnetization that is characteristic of all ferromagnetic substances.

Magnetic Fossils

It had been vaguely known since the middle of the 19th century that some rocks and baked earth and earthen vessels apparently retain the direction of past geomagnetic fields. However, it was not until the 1920's that this knowledge was utilized to any extent. The first studies of rock magnetism were made in Germany, France and Japan. It was in the early 1940's that J. G. Königsberger, E. and O. Thellier, and T. Nagata began to probe the fundamental question of why magnetic fossils occur in rocks (14). Then later in the decade, L. Néel of France, a theoretical physicist, became deeply interested in rock magnetism and laid the theoretical foundations for this science.

As fossil magnetism played a crucial role in the dramatic revival of the theory of continental drift in the 1950's, and more-over, has revealed fascinating facts about the past geomagnetic field, we propose to study its mechanism in this chapter. First, how does a rock acquire its fossil magnetism, its remanent magnetization?

Volcanic Rocks: Ideal Permanent Magnets

The magnetization of a ferromagnetic substance by an applied field is illustrated in Figure 4-2. If we start applying a magnetic field to a ferromagnetic substance of zero magnetization, the magnetization increases with the strength of the applied field along the broken line A. Eventually it stops increasing; the magnetization at this point (J_s) is called "saturated magnetization."

Now decrease the strength of the applied field. Magnetization falls off, not along curve A but along curve B. Even when the field strength is reduced to zero, the substance still retains some magnetization (J_r in Figure 4-2). This is the *remanent magnetization*.

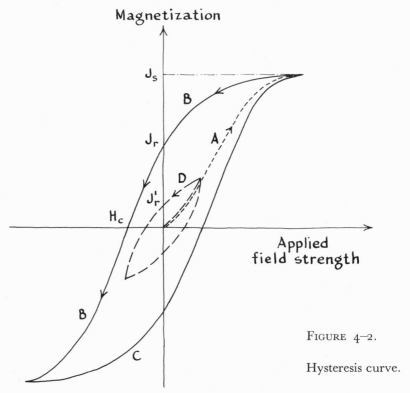

FIGURE 4–2.

Hysteresis curve.

A ferromagnetic substance with a large remanent magnetization makes a good permanent magnet.

Now apply a field of the opposite polarity. Magnetization decreases along the curve B; it becomes zero when the applied field strength reaches a certain value (H_c in Figure 4-2). H_c—the value of the negative field at which complete demagnetization occurs— is called *coercive force*. The resistance of the remanent magnetism of a substance against a negative field can be measured by the coercive force. A large coercive force, which means great resistance, is also a prerequisite for a stable permanent magnet.

When the negative field strength is increased still further, the substance is magnetized in the opposite (negative) direction along curve B, eventually reaching saturation. When the negative field is decreased, magnetization follows curve C (note that it is the inverted form of curve B). The curves in Figure 4-2 constitute the *hysteresis curve* that is characteristic of ferromagnetic substances.

In order to obtain remanent magnetization, it is not necessary to raise an applied field to the saturation point; application of a weak field followed by its reduction to zero will suffice; in that case, magnetization follows curve D, and *some* remanent magnetization (J_r') is obtained. This is precisely what happens when a ferromagnetic substance is subjected to the weak geomagnetic field.

A remanent magnetization possessed by a rock in its natural site is called its *natural remanent magnetization*. A volcanic rock containing ferromagnetic substances has a natural remanent magnetization, acquired when the rock was formed from lava under the influence of the prevailing geomagnetic field.

If a magnetic field is applied to a demagnetized volcanic rock in a laboratory at room temperature, remanent magnetization thus obtained is one hundred to one thousand times weaker than its natural remanent magnetization. What does this mean? Possibly, that at the time of formation of the volcanic rock, the geomagnetic field was a hundred to a thousand times stronger than it is now. However, this was found not to be so, for newly erupted volcanic rocks also possess the strong natural remanent magnetization.

Rock magnetists then conjectured that volcanic rocks acquire a strong natural remanent magnetization because they are formed under a special condition; *they are cooled from a high temperature* in the geomagnetic field, though the field is a weak one. To reproduce natural remanent magnetization in a laboratory, rocks and baked earth were heated to high temperatures and then cooled in a weak magnetic field. These experiments confirmed the conjecture. Such remanent magnetization is called *thermoremanent magnetization*.

Strikingly, in the process of cooling from high temperatures, a ferromagnetic substance acquires its thermoremanent magnetization mostly as it cools through the Curie point (see page 113). Suppose a rock has a Curie point of 500°C. Experiments have shown that nearly all the thermoremanent magnetization is acquired while the rock sample passes through 500°C to about 450°C. Below 450°C, only very little thermoremanent magnetization is acquired.

Thermoremanent magnetization is not only strong; it is also stable. Ordinary remanent magnetization, acquired in a field of 1 gauss, would be completely demagnetized by a slight increase in temperature, or by the application of an alternating field (a field periodically changing its polarity) of about 1 gauss. *Thermo-remanent* magnetization acquired in a field of 1 gauss will *not* be demagnetized by an alternating field of several hundred gauss, nor will temperature have much effect until it is raised almost to the Curie point. Thus thermoremanent magnetization has strong resistance or a large coercive force. Since a large remanent magnetization and a large coercive force make a good permanent magnet, volcanic rocks and baked earth are indeed ideal permanent magnets. That is why they can remain unaffected by all the disturbances exerted in the course of the long geologic periods, and be fossils of past geomagnetic fields.

A Chance for Ferromagnetic Control

How is it that a strong stable remanent magnetization is acquired only when a ferromagnetic substance cools through the Curie point? This question is fundamental to an understanding of rock magnetism, and many hypotheses for it have been proposed. As we shall soon see, rocks owe their ferromagnetism to several minerals, and one explanation may not fit all of them. However, the following analogy may illustrate the general idea.

Consider a volcanic rock which has just erupted as lava. Its temperature is, of course, above the Curie point, and therefore, there is no ferromagnetic order among the magnetic moments of its atoms; the magnetic moments are oriented quite at random, as in paramagnetic substances. Although exposed to the geomagnetic field, the lava acquires no strong magnetization. The lava gradually cools, its temperature passing through the Curie point. From this point, ferromagnetic control overcomes the thermal energy of atoms; the magnetic moments of atoms begin to align parallel to one another in the direction of the geomagnetic field. It is as if a large crowd of people, hitherto uncontrolled, suddenly lined up, facing one direction. Meanwhile the temperature falls lower and lower. Each atom is now, so to speak, arm-in-arm with its neighbors, facing one direction. The atoms lack the thermal energy to reverse their direction or to move at all. Having lost their freedom they are at the mercy of magnetism. Thus the remanent magnetization of the cooled lava is very stable. It can be demagnetized now only by raising the temperature to the Curie point—giving the thermal energy to the atoms—or by applying to it a particularly strong magnetic field.

What Minerals Are Responsible for the Ferromagnetic Property of Rocks?

Not all the constituent minerals of a rock are ferromagnetic. A rock generally contains some fine grains of iron oxide such as magnetite and hematite. It is these two minerals, although they may constitute only about 1% of a rock, that are responsible for its ferromagnetic properties. As the amount is so small, the magnetic intensity of a rock on the whole is far weaker than that of iron, but measured with a sensitive instrument, it exhibits the magnetization curve and the remanent magnetization characteristic of ferromagnetic substances.

Magnetite, as the name implies, is highly magnetic, and when brought near a magnet, is strongly attracted. Hematite is about one hundred times weaker in magnetic intensity. When placed close to a very strong magnet, it makes a just perceptible movement.

Pure magnetite (Fe_3O_4) and hematite (Fe_2O_3) are composed of iron and oxygen atoms only. However, in the magnetite and hematite found in natural rocks, a number of iron atoms are often replaced by atoms of other minerals, notably titanium. Therefore, these iron oxides are referred to as titano-magnetite or titano-hematite.

Titano-magnetite and titano-hematite are non-metallic ferromagnetics. The mechanism underlying their ferromagnetism differs from that of metals such as iron which was explained in Chapter 3 (page 112). It is only recently, in the 1950's, that the ferromagnetism of non-metallic substances was explained according to their atomic structures. There was a reason for this delay.

A Group of Perverse Atoms

Although man knew magnetite and hematite as ferromagnetic substances before he did iron or nickel, the study of ferromagnetism was, until recently, almost limited to iron, nickel, cobalt and their alloys. This was inevitable, as the usual approach in solid state physics is to start with the study of simple substances. Moreover, metallic ferromagnetic substances, particularly those with weak coercive force, were widely used—as electric magnets and the cores of electric coils for example—in engineering and industry. There, however, complaints about metallic ferromagnetics mounted. Because of their high electric conductivity, large amounts of induction current flow when the frequency of the current is very high; and the larger the induction current, the greater the energy loss. During the 1930's and 1940's, ferromagnetic substances of low electrical conductivity, non-metallic ferromagnetics, became important subjects of research in solid state physics.

Néel was a pioneer in research on non-metallic ferromagnetic materials (14). He believed that among the atoms of iron oxide, there exists a strong interaction comparable to the interaction of atoms in metallic iron. In iron oxide, the interaction is mediated by the large oxygen atom which stands between two iron atoms. Moreover, the magnetic moments of two neighboring iron atoms, with the oxygen atom between, tend to take opposite directions. If the atoms in a ferromagnetic metal resemble a group regimentally aligned without individuality, the atoms in iron oxide resemble a group of utterly perverse beings. However, since every iron atom is equally perverse, there emerges a kind of order; the magnetic moment of *every* iron atom is opposite to that of its neighbor.

The astute reader will have assumed that if the magnetic moment of every atom is opposite to that of its neighbor, the magnetic moments will annul one another, with the result that the substance as a whole would not be ferromagnetic. This is correct. But with magnetite, there is a reason why this does not happen. The chemical formula for magnetite Fe_3O_4 can be rewritten as $FeO \cdot Fe_2O_3$. In a mineral, atoms are arranged in a geometrical lattice-work, and there are two sub-lattices, say A and B, in magnetite. One half of the iron atoms of Fe_2O_3 belong to sub-lattice A. The remaining half of the iron atoms of Fe_2O_3 and all the iron atoms of FeO belong to sub-lattice B. This is schematized in Figure 4-3. The magnetic

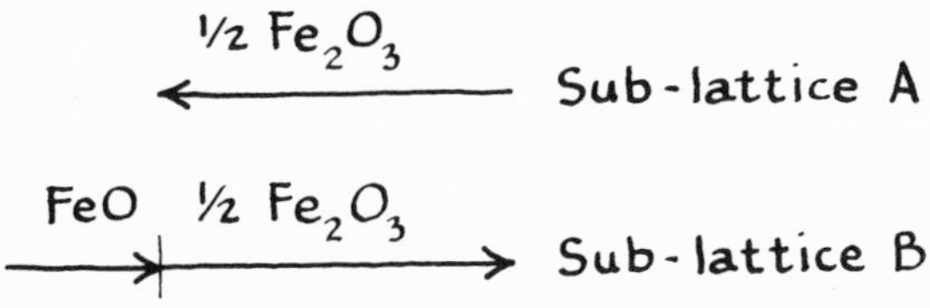

$$\tfrac{1}{2} \, Fe_2O_3$$
$$\longleftarrow \quad \text{Sub-lattice A}$$

FIGURE 4-3.

$$FeO \quad \tfrac{1}{2} \, Fe_2O_3$$
$$\longrightarrow\!\!|\!\longrightarrow \quad \text{Sub-lattice B}$$

moments of the iron atoms of Fe_2O_3 in A and B are equal in magnetic intensity but they are oriented in opposite directions. Therefore, they cancel each other. However, the magnetic moment of FeO survives, giving a magnetic moment to magnetite as a whole.

Néel showed that this applies not only to magnetite, but also to all ferromagnetic oxides known as ferrite. He named this type of ferromagnetism *ferrimagnetism* to distinguish it from the ferromagnetism of metals. In the case of hematite, the two antiparallel magnetic moments are of almost equal intensity; consequently, the magnetic moment of hematite as a whole is very weak. Nevertheless hematite plays an important role in rock magnetism (see page 147).

Almost all of Néel's theories have been either borne out by experiments or supported by more basic theories.

Was the South Once North?

Studies such as we have described, studies of rock magnetism, were basic to the development of paleomagnetism, the study of the history of the geomagnetic field through measurement of the directions of the remanent magnetization of rocks and baked earth and earthenware.

Up to the early 1950's, paleomagnetic studies were mostly confined to comparatively recent geologic periods such as the Quaternary and the Tertiary. A principal finding from these studies has been that the earth's field, despite the secular variations reported in the prior chapter, seems to have maintained a stable dipolar field since about the Tertiary period: the earth's field, then, has been more or less stable for about 60 million years.

What was puzzling to paleomagnetists, however, was that some rocks were discovered to be magnetized in a direction quite the reverse of that of the present geomagnetic field. The first of these reversely magnetized rocks was discovered in 1906 in France; since then, others have been found in areas as widely separated as Japan, Spitsbergen, Greenland, the Faroe Islands, England, Australia, Germany, South Africa, Iceland, the United States and the Soviet Union. Now we know that there are as many reversely-magnetized rocks as there are normally-magnetized ones, and this is a significant fact.

A possible explanation of reversed magnetization was that, at the time of the magnetization of these rocks, the direction of the geomagnetic field was the reverse of the present; the present North Pole was then the South Pole and vice versa. Furthermore, judging from the age of reversely-magnetized rocks, the earth's field must have reversed itself not only once, but many times since the

Tertiary. This speculation about reversal of the field startled many a cautious scientist. Then, when Blackett proposed his hypothesis on the origin of the geomagnetic field, the implications became positively fascinating. According to Blackett, remember, the rotation of the earth causes the geomagnetic field. If Blackett's idea and the field-reversal hypothesis were both correct, it would mean that, during the reverse-polarity epoch, the earth rotated in the opposite direction. Not only were north and south reversed, but east and west as well. Could the earth rotate in one direction and then the other? That would be absurd to suggest.

Then, in 1949, an American rock magnetist, John Graham, lighted on an idea and wrote Néel asking whether the latter could think of any mechanism within a rock that might reverse its magnetization without the magnetic field of the earth itself having been reversed. Three years later, in an article in the *Annale de Géophysiques*, Néel wrote that he considered it theoretically possible for a rock to acquire reverse magnetization through its own internal mechanisms (*14*).

The Power of Prediction

Néel pointed out four possibilities for self-reversal. Néel's third and fourth suggestions concern chemical changes which might, during long geological periods, account for change in the direction of magnetization. Interesting as these possibilities are, they are not pertinent to our immediate question. We therefore concentrate on Néel's first two suggestions.

Néel's first suggestion concerns a peculiar case of ferrimagnetism (see page 141). In a ferrimagnetic substance with two sublattices A and B, the magnetic moments of all the magnetic atoms in lattice A are oppositely directed to those of lattice B. Suppose the magnetic moments of lattice A and lattice B vary differently with temperature in such a way that the magnetic moment of A is larger than that of B over certain temperature range but that it is the reverse elsewhere (see Figure 4-5). The overall magnetization, which is the joint effect of the magnetic moments of A and B, can then take either normal or reverse polarity depending on whether A is larger than B or not. When Néel set forth this possibility, no one knew whether such a substance actually existed, let alone whether it occurred in rocks.

FIGURE 4–4.

L. Néel.

Later, in 1953, E. W. Gorter of Holland synthesized a substance that behaves in this way (14); an iron oxide containing lithium and chromium reversed its magnetization between room temperature and the Curie point, just as Néel had predicted. Discovery of similar substances soon followed. However, as these substances were all synthetic, they could not account for the reversed remanent magnetization found in natural rocks.

Néel's second suggestion involved what is called magnetostatic interaction between two phases. It assumes that two ferromagnetic substances, A and B, with different Curie points, are contained in a rock. Suppose this rock cools from a high temperature in the geomagnetic field. Substance A with the higher Curie point would acquire thermoremanent magnetization in the field direction. The intensity of the thermoremanent magnetization would increase as the rock cools (see Figure 4-6, curve A). As the temperature falls to the Curie point of substance B, the latter would acquire its

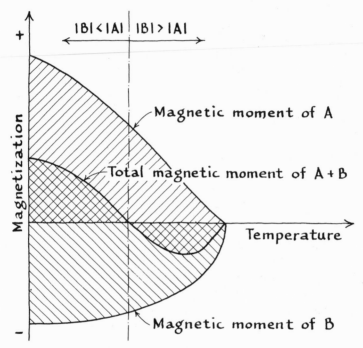

FIGURE 4-5.

magnetization; however, B is subject not only to the earth's field, H_e, in Figure 4-7, but also to the field H_i of the substance A which has already become a permanent magnet. If the substance B is located at the position shown in Figure 4-7, the direction of H_i is opposite to that of H_e in B. Under certain conditions the effect of H_i can be stronger than that of H_e so that the substance B is reversely magnetized. As the temperature falls, the magnetization of substance B would follow curve B in Figure 4-6. It is possible

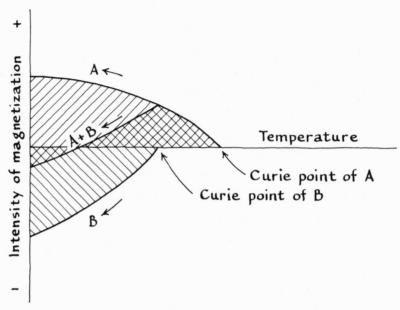

FIGURE 4-6.

that the reverse magnetization of substance B becomes stronger
than the normal magnetization of substance A under favorable
conditions. That is, the resultant magnetization of the whole rock—
the combined effect of A and B—follows the curve A+B in
Figure 4-6, reversing its direction during cooling.

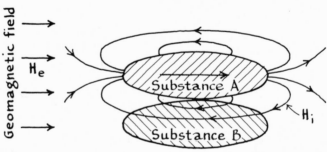

FIGURE 4-7.

Self-Reversal Occurred in a Natural Rock

In the same year, 1952, a phenomenon was discovered in the rock magnetism laboratory at the University of Tokyo; a pumice from the volcanic area of Haruna, Japan, was found to acquire thermo-remanent magnetization in a direction opposite to that of the applied field. The phenomenon was named "reverse thermorem-anent magnetization." This discovery was made before the Jap-anese rock magnetists became aware of Néel's article published a few months earlier in France. We also thought that the phenom-enon might have resulted from an interaction of two substances with different Curie points, and proceeded to analyze the pumice chemically. Two such substances were found. One was a titano-magnetite with a Curie point around 500°C; the other was an ilmenite-hematite (a solid solution of ilmenite ($FeTiO_3$) and hematite (Fe_2O_3)) with a Curie point around 250°C.

The self-reversing thermoremanent magnetization of the Haruna pumice was soon acclaimed as a proof of Néel's prediction. Closer, more careful study revealed, however—somewhat to our embarrassment—that the mechanism of the self-reversal of the Haruna rock actually differed from what Néel and we had pre-dicted. S. Uyeda (1958) found that the rock still showed self-reversal, even much more strongly, after we had removed the titano-magnetite from the Haruna pumice, leaving only the ilmenite-hematite (14). Thus the theory of interaction between two substances did not apply to the reversal of the Haruna rock. The ilmenite-hematite alone was responsible. This substance was indeed mysterious. As we have mentioned, the magnetic intensity of hematite is much weaker than that of magnetite. Yet the hematite in the pumice, containing a great deal of titanium, showed a high magnetic intensity, comparable to that of magnetite.

The mystery was in the main solved around 1960. It was found that a reverse thermoremanent magnetization occurs in an ilmenite-hematite when the ratio of ilmenite to hematite is around 1. This phenomenon is now explained as the result of a peculiar configuration of iron and titanium atoms in the crystal, but the explanation of the mechanism, which is extremely complex, is beyond the scope of this book.

The Earth's Field Does Seem to Have Reversed

Does the self-reversing mechanism of Haruna pumice explain the reversed remanent magnetization of natural rocks? For some time after the discovery of the Haruna rock, many rock magnetists thought that self-reversal accounted for all natural reverse magnetization. During 1953 and 1954, experiments sought to verify the point. Out of fifty or sixty cases of natural reverse magnetization, only a few were found to be self-reversals. The rest supported the hypothesis that the geomagnetic field reversed itself.

More compelling evidence for this hypothesis was to come. Sometimes a dyke of igneous rocks intrudes into a stratum long after the initial formation of the latter. The intrusion is not unusual, but when the direction of the remanent magnetization of the original stratum and that of the dyke were measured, an interesting fact emerged. Figure 4-8 shows the result obtained by Y. Kato and

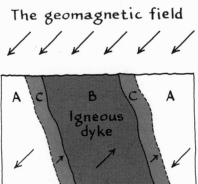

FIGURE 4-8.

others (*8, 14*). The part of the stratum far removed from the dyke, A, showed normal magnetization, but the dyke, B, and the part adjoining and probably baked by the dyke, C, were found to be magnetized in a reverse direction.

Perhaps the geomagnetic field had a normal polarity when stratum A was formed, but a reverse polarity when dyke B intruded. Thus, reverse magnetization was acquired not only by the dyke B, but also by C, because this part of the stratum, having been heated above the Curie point by contact with the hot lava of the dyke, lost the old remanent magnetization and acquired a new thermo-remanent magnetization in the direction of the geomagnetic field at the time of intrusion. Though there is a chance that the dyke B possessed the self-reversal property, it is highly improbable that stratum A, differing in mineral composition, also possessed it.

There is other evidence in support of the field reversal hypothesis. K. Momose and others reported that certain Tertiary strata in the central part of Japan show reverse remanent magnetization

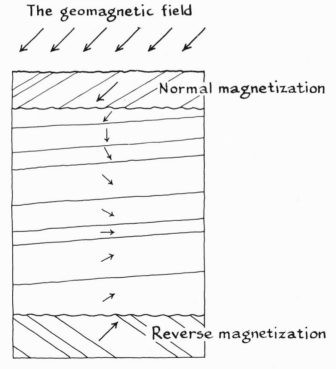

FIGURE 4-9.

though the strata above them are magnetized in the normal direction. This itself is not so rare, but the rocks between the strata of normal magnetization and those of reverse magnetization show *intermediate directions* of magnetization, as shown in Figure 4-9. Could this have been caused by some intrinsic property of the rocks? Does it not rather demonstrate that the geomagnetic field *rotated* at the time of reversal? As supporting evidence mounted, the field reversal hypothesis was more widely accepted as valid.

Once the idea of geomagnetic field reversal is accepted, the question arises: when and how often has it happened? Fortunately, since the radioactive method of dating rocks is available, the question can be settled by dating all rocks that are reversely magnetized. R. Doell, A. Cox, B. Dalrymple and their colleagues at

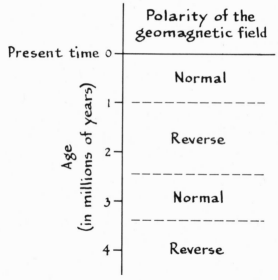

FIGURE 4-10.

Polarity of the geomagnetic field in the past. (Adapted from figure 3 in Reversals of the earth's magnetic field, by A. Cox, R. Doell, and B. Dalrymple, in *Science*, *144*, *1537–1543*, *1964*, with permission of the authors.)

the U.S. Geological Survey have set about this task. Although the work is in its initial stage, they have already reported a periodic reversal of the earth's field (3). In Figure 4-10, the vertical axis represents the absolute geologic age, and the direction of remanent magnetization, normal or reverse, is given. According to Figure 4-10, there have been three major reversals in the last 3.4 million years, and the duration of the normal-polarity and that of the reverse-polarity epochs are about equal. Moreover, within the epochs, which are of the order of one million years, there are short intervals of about ten thousand years when the polarity was the reverse of the generally prevailing one.

If field reversal actually occurred, its evidence must be found in rocks all over the globe. The pattern shown in Figure 4-10 is chiefly based on rocks from the western part of the United States, Alaska and Hawaii. Attempts to corroborate these findings with rocks from all over the world are under way.

What Reversed the Geomagnetic Field?

Periodic reversal of the geomagnetic field, assuming it has actually occurred, must be taken into account in any theory on the origin of the geomagnetic field. The hypothesis that the geomagnetic field owes its direct origin to the rotation of the earth is not valid, for it is highly unlikely that the direction of rotation changed with every field reversal.

The dynamo theory has advantages over other theories, for it has been demonstrated that a field of either polarity, normal or reverse, could originate from the postulated fluid motion. The dynamo theory to date, which assumes a steady-state dynamo and only discusses its stability, cannot yet explain the concrete processes of field reversal. To study the time-dependent behavior of the dynamo, the only method pursued so far is to study a simple model (such as the disc dynamo mentioned in the previous chapter), and to infer the mechanism of the earth dynamo therefrom. For instance, it has been proved by T. Rikitake (15) as well as others that periodic reversal is possible when there is an interaction between two disc dynamos. Still, the mechanism of such a model must differ considerably from what is actually happening in the earth's core. Development of a comprehensive theory of the self-exciting dynamo that can explain the geomagnetic field reversal is one of the greatest tasks that modern geophysics faces today.

Concluding Remarks

This chapter has mainly dealt with the mechanism of fossil magnetism and the interesting facts it revealed about the past geomagnetic field. Since the stability of the thermoremanent magnetization of a rock has been generally confirmed, it can reliably be used as a fossil compass indicating the direction—declination and inclination—of the past geomagnetic field. The next chapter will show how quantitative data of this kind shed a new light on, and gave a sounder basis to, the discussion of possible continental movement in the past.

5

Revival of the Continental Drift Theory

Tracing Back the Whole Geologic Past

In the early 1950's, the main topics of interest in paleomagnetism were the possibility of the geomagnetic field reversal and self-reversing thermoremanent magnetization. French and Japanese scientists were active in this field. Then, rather suddenly, paleomagnetic works began to appear from England as well. Blackett of London University had become interested in the reversal of rock magnetism during his "negative experiment." Runcorn who was then at Cambridge University (he is now at the University of Newcastle-upon-Tyne) had become independently interested in the same problem when a reverse magnetization was discovered in lava from Iceland.

At first, most works by English paleomagnetists seemed somewhat rudimentary, for they simply reported that such and such a stratum had natural remanent magnetization in such and such direction. However, the total volume of their work increased rapidly. We rock magnetists in Japan began to suspect that their paleomagnetism might develop in some new direction, different from our own.

To begin with, a remarkable fact about the English work was

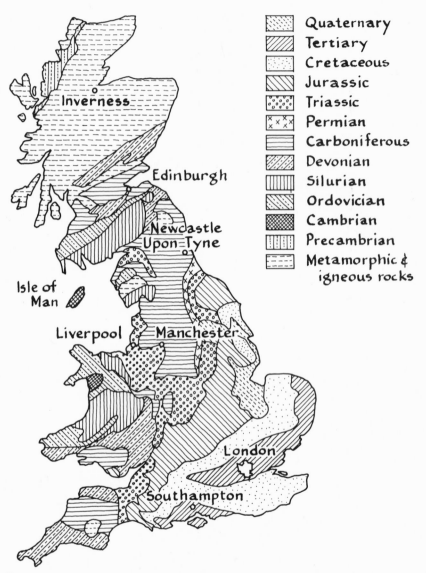

▨	Quaternary
▨	Tertiary
⬚	Cretaceous
◩	Jurassic
⬚	Triassic
⊠	Permian
▤	Carboniferous
▨	Devonian
▥	Silurian
▨	Ordovician
▨	Cambrian
▥	Precambrian
⬚	Metamorphic & igneous rocks

FIGURE 5-1.

Geologic map of England.

that they extended the application of the paleomagnetic method
to the whole of geologic history instead of only as far back as the
Tertiary. For this purpose, England has a great geologic advantage, for, amazingly enough, strata of every geologic period from
the Precambrian to the Quaternary, are to be found on this small
island, preserved relatively undisturbed, like annual rings of a very
old tree (see Figure 5-1). In Japan, pre-Tertiary strata have generally been folded and refolded by violent orogeneses, dislocated
by faults, and, in short, quite deformed out of their original shapes.
Needless to say, the paleomagnetic inference of the past geomagnetic field direction rests on the assumption that the stratum itself
has not changed its orientation since its formation. If it has tilted
because of geologic activities, we must know how it has tilted; that
is, we must be able to infer its original orientation. This is almost
impossible with the older strata of Japan because of the violent
disturbances which seem to be the fate of all island arcs around
the Pacific.

In England, magnetists made the most of their geologic advantage; if the paleomagnetic method could be applied to strata a
million years old, they argued, why not to those a billion years old?

In a volcanic country like Japan, strata are often composed of
igneous rocks whose thermoremanent magnetization is very strong.
In England, strata are mostly composed of sedimentary rocks whose
remanent magnetization, as we shall see in the next section, is often
very weak. Fortunately, however, Blackett's ultra-sensitive astatic
magnetometer, originally designed for the "negative experiment,"
was now at hand for the measurement of weak remanent magnetism.

Depositional Remanent Magnetization

Rocks at the earth's surface are weathered by rain and wind, and
the loosened fragments are carried by rivers to lakes and the sea
(erosion). These rock particles are deposited at the bottom of the
water (deposition). The products of deposition, the sedimentary
layers, eventually become *sedimentary rocks* by chemical action and

physical forces such as pressure. The process is described by an ancient poet in the national anthem of Japan:

> May the Emperor reign
> For thousands of years
> *Till pebbles become rocks*
> And rocks gather moss

Today, a substantial portion of the land surface is covered by sedimentary rocks. In rock magnetism, however, research was first concentrated on igneous rocks because they presented the interesting phenomenon of thermoremanent magnetization, and this magnetization, being strong, was easy to measure. Still, there would have been some real advantage in improving the measurement technique and thus making sedimentary rocks available for paleomagnetic use. Sedimentary rocks are ideal for an exhaustive study of the geologic past, for deposition is a continuous process while igneous activities are sporadic.

Unlike igneous rocks, sedimentary rocks have not been cooled from a high temperature in the geomagnetic field. Therefore, they have no thermoremanent magnetization. What, then, is the mechanism by which sedimentary rocks retain the direction of the geomagnetic field prevailing at the time of deposition?

Consider the experiment illustrated in Figure 5-2. Tiny magnetic needles are mixed with grains of sand and thrown little by little into water in a container. A magnetic field is applied by placing one end of a large magnet close to the bottom of the container. The deposits which accumulate there are, so to speak, laboratory reproductions of sedimentary rocks formed in the geomagnetic field. Upon measurement, we find that the direction of the remanent magnetization acquired by these deposits coincides with the direction of the magnetic field applied during deposition. Why? The small magnets, while falling through the water, orient themselves in the applied field direction before settling at the bottom. Once deposited, they are so firmly fixed among sand grains that their orientation does not change, even upon the removal of the field. This is how they acquire remanent magnetization in the direction of the applied field.

Rock particles eroded from mountains contain grains of magnetite and hematite which are in fact small magnets, for they must have acquired thermoremanent magnetization when the original

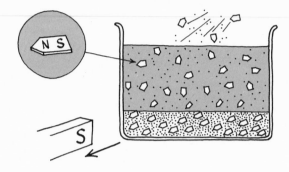

FIGURE 5–2.

Experiment illustrating the acquisition of depositional remanent magnetization.

igneous rock was formed. Therefore, during deposition, these magnetized particles act just as do the small magnets in the experiment; they orient themselves in the geomagnetic field direction. The natural remanent magnetization of sedimentary rocks thus acquired is called *depositional remanent magnetization.*

Of course, the magnetic moments of these ferromagnetic grains are far weaker than those of magnetic needles used in the experiment. The earth's field itself is not very strong. Therefore, the tendency to align in its field direction is actually very weak; in fact, only a small portion of magnetized grains orient themselves in this way. The depositional remanent magnetization of sedimentary rocks is about a hundred times weaker than the thermoremanent magnetization of igneous rocks. Its measurement does pose difficulties.

Problems of Depositional Remanent Magnetization

A closer study of the depositional remanent magnetization revealed that it is not only weak but also often inconvenient for paleomagnetic purposes. The trouble is that, although its direction on the horizontal plane—declination—coincides with that of the applied field, its direction on the vertical plane—inclination—tends to be less than that of the applied field; for instance, if the inclination of the applied field is 70°, that of the depositional remanent magnetization tends to be about 50° (see Figure 5-3). If the deposition

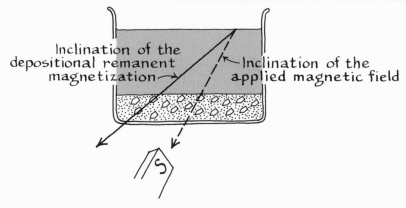

FIGURE 5-3.

Inclination error.

is made in running water or on a sloped surface, there is further deviation in inclination. This phenomenon, called "inclination error," is a thorn in the side of paleomagnetists. However, some paleomagnetists hold that in some cases, igneous rocks and sedimentary rocks of the same age match well in inclination and that therefore the effect of this inclination error is not so serious.

What causes the inclination error? According to R. King, D. H. Griffith, A. Rees and others at Birmingham University (8, 14), the ferromagnetic mineral grains roll slightly as they settle on the deposition surface. Statistically, this rolling results in the diminution of the inclination value.

Another defect of depositional remanent magnetization is its instability. Paleomagnetism, as we have often noted, rests on the assumption that the remanent magnetization acquired at the time of deposition is faithfully preserved. Some sedimentary rock samples, left in the laboratory, change their magnetization directions within a few months or even a few days. With such fickle short-term behavior, there is no knowing what changes the sediments have undergone in the long geologic past.

This instability is due to the acquisition, in addition to the original magnetization, of a new remanent magnetization in the direction of the geomagnetic field to which the rock has been exposed after the acquisition of the original magnetization. This is called *viscous* or *secondary* remanent magnetization.

Further study showed that this secondary remanent magnetization can occur in all kinds of rocks. The problem lies in its intensity compared to that of the original remanent magnetization. If the secondary magnetization is negligibly small, there is no trouble. If it is large, it obscures the original magnetization. The intensity of secondary remanent magnetization differs with rocks and we do not yet know what determines this difference. Its effect is particularly conspicuous in sedimentary rocks because the original magnetization is so weak. At any rate, this instability is fatal for paleomagnetism.

Thus, it became imperative to develop a method for discriminating between usable and unusable rocks. Various methods have been devised, and now it is an established rule to put a rock sample through a series of experiments to ensure the stability of its remanent magnetization before using it for paleomagnetic purposes.

Graham Test

Such experiments on stability are after all artificial laboratory devices. There is no direct way of determining whether a rock, exposed to all the changes of nature for millions or billions of years, has really preserved faithfully the original direction of its remanent magnetization. At least, more cautious scientists are sceptical on this point. Therefore, we will introduce one test called the Graham test which may be acceptable. J. Graham is an American rock magnetist with remarkable resourcefulness and ideas.

Look at Figure 5-4. Conglomerate, a kind of sedimentary rock,

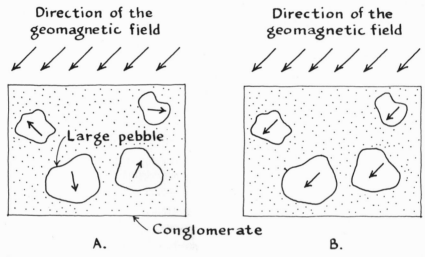

FIGURE 5–4.

Graham test.

contains many large pebbles. These are actually fragments of rocks carried away by a rapid stream, deposited, and cemented together. These pebbles, because of their size, cannot possibly have oriented themselves in the geomagnetic field direction at the time of deposition; the effect of gravity on their relatively large mass was greater than the effect of the magnetic field. Thus they must have settled down in quite random directions. Now, if we measure the magnetization directions of these pebbles in conglomerate, they are sometimes random as in Figure 5-4A and sometimes aligned in the present geomagnetic field direction (Figure 5-4B). Obviously, in the case of (A), the remanent magnetization has been stable throughout the geologic past, and in the case of (B) has changed because of strong secondary remanent magnetization. Therefore, if the parent rock of the pebbles in case A can be found, the remanent magnetization of that rock can be considered as having been stable and is therefore qualified for paleomagnetic use. A number of such instances have been found. Therefore, we can at least reject the extremely pessimistic view that all rock magnetization is unstable.

Chemical Remanent Magnetization

Having considered the thermoremanent magnetization of igneous rocks and the depositional remanent magnetization of sedimentary rocks, we now turn to *chemical remanent magnetization*, which is the property of a third kind of rock, namely, *metamorphic* rocks. A metamorphic rock is formed when in the course of geologic time, igneous or sedimentary rocks undergo transformation because of chemical changes or physical forces.

If, in the process of the metamorphic transformation (metamorphism), the temperature of the rock were raised—as it often is—above the Curie point, then the metamorphic rock would acquire thermoremanent magnetization and thus reflect the geomagnetic field direction at the time of metamorphism. This is easy to understand and reasonable to accept.

In some cases, however, long after the formation of a rock, a ferromagnetic mineral is newly produced within the rock by some chemical changes, at a temperature far below the Curie point. For instance, underground water containing iron may flow through sedimentary rocks and precipitate iron oxides or hydroxides; such hydrated iron may change into hematite when it is dehydrated; or iron contained in non-ferromagnetic minerals may turn into ferromagnetic oxides through various chemical processes. In the light of what we have considered so far in this study, no strong thermoremanent magnetization would be acquired in these cases, since the temperature is far below the Curie point. However, as data were accumulated, it was found that rocks in which ferromagnetic minerals were evidently produced after deposition often showed fairly strong remanent magnetization.

This makes us suspect that some kind of remanent magnetization may be acquired even at a low temperature when a ferromagnetic mineral is newly produced under the influence of a magnetic field. In 1958, this problem was taken up for intensive study by K. Kobayashi and others at the University of Tokyo (8, 14). The experiment devised for the study of this problem was a simple one. Magnetite was chemically synthesized at a temperature around 300°C in the geomagnetic field (at this temperature, which is about 250°C below the Curie point for magnetite, hardly any thermoremanent magnetization would be acquired). As expected, a fairly strong remanent magnetization was produced. This phenomenon was interpreted theoretically as follows. The synthesis of a ferromagnetic substance such as magnetite implies that the magnetic moments of atoms which had previously been random become aligned in one direction. Consider the contrast. In the case of thermoremanent magnetization, the magnetic moments of atoms succumb to ferromagnetic control at the Curie point of the substance when the exchange interaction becomes stronger than the thermal energy of atoms. In the case of chemical remanent magnetization, it is the rearrangement of atoms through chemical changes or crystallization that brings about the alignment. Thus the transition to alignment occurs in both cases although the causes differ. Therefore if, at the time of transition, the rock sample is under the influence of a magnetic field—even a weak one—the magnetic moments of atoms align in the field direction.

Just as this research was concluded in Tokyo, a paper entitled "The process of magnetization by chemical changes," was published by G. Haigh of London University (8, 14). His findings coincided exactly with the conclusions of the Tokyo University group. Even the details of the supporting experiments were strikingly similar.

Such coincidences of independent work on the same theme

occur fairly often in academic circles, particularly when, as in this century, important problems and work on them are rather quickly communicated to all parts of the world. Whenever such a coincidence occurs, we cannot help feeling that men, no matter where they live, think alike. At the same time, it brings home to us the fact that even geophysics which deals with millions and billions of years is no longer the leisurely science that it used to be, but has become an arena for keen and healthy competition.

Thus, chemical remanent magnetization came to be recognized as a definite phenomenon. Its significance turned out to be greater than was first imagined, for it is possible that the remanent magnetization of red sandstone, a kind of sedimentary rock often used in paleomagnetism, is of chemical rather than depositional origin. If so, we need not worry about its inclination error.

Has England Rotated Clockwise?

The first remarkable achievement of English paleomagnetists resulted from the study of the Triassic strata (see page 154) conducted by the London University group. Triassic layers, composed of red sandstones, are found in various parts of England. Around 1953, the natural remanent magnetization directions of these rocks were extensively measured. The result was that about half the rock samples showed a mean declination of N.29°E (i.e., 29° East of the present North) and a mean inclination of 34° downward, while the remaining half showed a mean declination of S.39°W and a mean inclination of 16° upward. In order to grasp the significance of this interesting result, we may adopt the following simple method of presenting paleomagnetic data.

Look at Figure 5-5A. From the center of the sphere (O), an arrow indicating the direction of the remanent magnetization is drawn till it intersects the surface of the sphere. The intersecting point is marked with a dot (·). If we look down on the sphere from directly above, it will look like Figure 5-5B. In this way, the remanent magnetization direction is expressed on a plane.

In projection, instead of simply looking down on the sphere

from above, we use the following procedure. Let O' be the point of contact between the sphere and the horizontal plane on which it is placed. The projected point P is determined in such a way

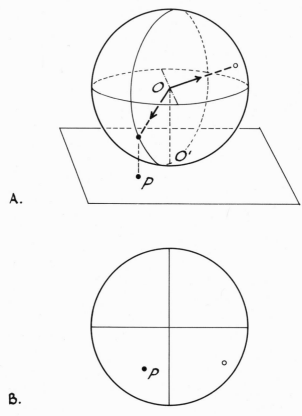

A.

B.

FIGURE 5–5.

Projective method of presenting paleomagnetic data.

that $O'P$ is equal in length to the arc formed by O' and the dot ($\cdot$). If the arrow points downward, the intersection of the arrow and the surface of the sphere would be in the lower hemisphere. If the arrow points upward, it would be in the upper hemisphere. In order to distinguish between the two, the intersecting point in the lower hemisphere is indicated by a dot (·), and that in the upper hemisphere by an open circle (∘).

The remanent magnetization directions of the Triassic red sandstones, presented by this projective method, are shown in Figure 5-6. In this figure, the cross (x) indicates the present geo-

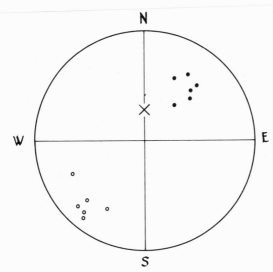

FIGURE 5–6.

Remanent magnetization directions of Triassic red sandstones in England, presented by the projective method. (Adapted from The remanent magnetism of sedimentary rock in Britain, by J. A. Clegg, M. Almond, and P. H. S. Stubbs, in *Phil. Mag. 45. 583–598, 1954,* with permission of the authors.)

magnetic field direction (as England is in the Northern Hemisphere, the present inclination dips downward). The dots and the open circles fall neatly into two groups and the difference between the average directions of the two is about 180°. This difference reflects the possible reversal in polarity which we mentioned in the previous chapter. What is more important here, we can see that the geomagnetic field direction in the Triassic deviates considerably from the present field direction. Compare Figure 5-6 with Figure 5-7 which shows the remanent magnetization directions of post-Tertiary rocks. In the latter case, the mean direction (apart from the reversal) is close to the present field direction. In contrast, the geomagnetic field direction in the Triassic (about 200 million years

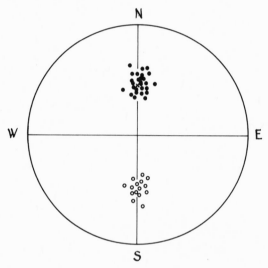

FIGURE 5–7.

A schematic drawing of the remanent magnetization directions of post-Tertiary rocks.

ago) diverges considerably (at least in England) from that which has prevailed from the Tertiary to the present.

This phenomenon was interpreted as follows. It is not that the geomagnetic field direction itself has changed from the Triassic to the present, but that England has rotated about 30° clockwise since the Triassic (see Figure 5-8). This explains the deflection in declination. As for inclination which was about 30° in the Triassic and is now 65°, it was suggested in explanation of this discrepancy

that England must have been situated at a lower latitude during
the Triassic, for as we know, inclination is 90° at the Poles and
decreases with latitude. If England was situated at a lower latitude, she must have had a warmer climate. Red sandstones generally occur in warm arid climate. Therefore, the inference that
England was situated at a lower latitude seems reasonable from a
paleoclimatological point of view as well.

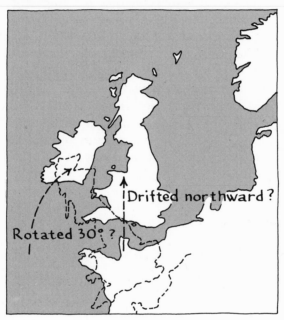

FIGURE 5-8.

Thus, all of a sudden, paleomagnetism reopened the question
of the possibility of continental drift. We rock magnetists in Tokyo
realized with great excitement the relevance that the work of the
English scientists had for the continental drift theory.

However, in 1954, when this argument was put forward by
paleomagnetists, the controversy over continental drift had been
dead for more than two decades. Only a few scientists showed any
interest: the majority dismissed it as merely the repetition of an
old story.

The Northward Drift of India

Yet, the English did not relax in their efforts. Within a short time, the magnetism of most of the strata in England was studied and recorded. Soon, the English scientists were extending their activities to India, Africa and Australia, and further to the United States and even Antarctica.

The second great achievement of the English paleomagnetists resulted from the exhaustive measurement, conducted by the London University group, of rocks in the Deccan Plateau of India. This vast plateau was formed from the Jurassic to the Tertiary period, by the eruption of basaltic lava. According to measurements (apart from the reversals in polarity), the mean inclination was 64° *upward* in the Jurassic, 60° *upward* in the Cretaceous, 26° *upward* in the early Tertiary, and only in the middle Tertiary, does it become 17° *downward*. How are we to account for this phenomenon? If a rock of a given period preserves an upward inclination,

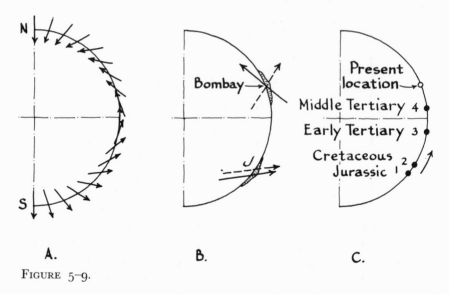

A.   B.   C.

FIGURE 5–9.

(Adapted from E. R. Deutsch, *J. of Alberta Society of Petroleum Geology*, 6, *1958, p. 155*, with permission of the author.)

it means that the sampling site of the rock was in the Southern Hemisphere. This is clear from Figure 5-9A which shows the inclination distribution of the dipolar field at various latitudes of the earth. Therefore, up to the early Tertiary period, the Indian peninsula must have been in the Southern Hemisphere.

By collating the inclination values of the remanent magnetization of Deccan rocks with those of the dipolar field in Figure 5-9A, we can determine more precisely the ancient geographical latitude of India. Look at Figure 5-9B. The dotted arrow represents the inclination value of the Jurassic basalt in India, while the solid arrow represents that of the dipolar field. In order to make the two arrows coincide, we must place India around point J in the Southern Hemisphere. To take Bombay as the reference point, this city which is today located at 19°N must have been located at 40°S in the Jurassic. By similar process, we can obtain the ancient latitudes of India for other geologic periods. The results are shown in Figure 5-9C. Since the Jurassic, the Indian peninsula has moved northward for about 7000 km, at the rate of a few centimeters per year.

The discerning reader will no doubt have realized that this northward movement of India is the very idea put forward, decades ago, by Wegener in the theory of continental drift. The idea of the drift, originally conceived in explanation of the distribution of lemurs, the origin of the Himalayas and the distribution of Permo-Carboniferous glaciers—the idea which had been so hotly contended and then forgotten—was now resurrected by the completely "independent evidence" of rock magnetism.

Up to now, the continental drift theory had chiefly rested on non-quantitative evidences such as the distribution of lemurs and earthworms. This time, the ancient latitude of India was mathematically determined from the direction of remanent magnetization. This was advantageous, since there is no room for subjective bias. For this reason, the new paleomagnetic evidence carried greater weight than the former evidences for cautious and sceptical scientists.

Refutation by the Newcastle Group

In the course of scientific progress, particularly in the initial stage of research, there is a frequent coincidence of independent work on the same theme. The attempt to revive the continental drift theory through paleomagnetism was a typical example. While the London University group was arriving at the above-mentioned conclusion, the Newcastle University group, led by Runcorn, was studying the same problem from another angle. The latter group objected cogently, as follows.

In inferring the drifting of the Indian peninsula and England from paleomagnetic evidence, the London group had made a tacit assumption that the earth's dipolar field has always remained unchanged. The assumption was that the geographic poles have always remained fixed and have always coincided approximately with the geomagnetic poles (see page 100). That is why a shallower inclination meant a lower latitude, and why an upward inclination signified a location in the Southern Hemisphere.

We must pause to consider whether this inference is really correct. First of all, what do the paleomagnetic data tell us directly, without any inference on our part? They tell us the ancient direction of the geomagnetic field at a given time and place (i.e., a locality whose *present* latitude and longitude we know). From this information alone, we can infer neither continental drift nor the ancient position of the geomagnetic poles. Thus we bring in our first assumption that the earth's field has always been the same dipolar field that it is now. This assumption is actually open to question (see page 186), but there is a strong paleomagnetic evidence that at least since the mid-Tertiary, the earth's field has been dipolar (see page 142). The dipolar field is a regular field (see Figure 5-9A). This is why we could determine precisely the ancient latitude of India from the inclination value of the geomagnetic field preserved in the Deccan rock samples. This latitude, however, is a latitude relative to the geomagnetic north pole, wherever it may have been. In other words, we are determining the distance between India and the geomagnetic north pole, or their positions

relative to one another. In order to conclude the northward drifting of the Indian peninsula from the change in its ancient latitude, we need the second assumption that the geomagnetic poles
have not moved relative to the *present* geographical poles. If the
geomagnetic poles have moved, the change in inclination can be
explained as due to this cause and the Indian peninsula need not
have moved at all. This second assumption that the geomagnetic
poles have always coincided approximately with the present geographical poles seems to be correct for the post-mid-Tertiary period.
Beyond that, the assumption is quite unproven. It is on the contrary
quite probable that the geomagnetic poles had shifted a great deal
prior to the Tertiary (see page 172). Therefore any change of position relative to the geomagnetic poles that have themselves moved
would not constitute a proof that England or the Indian peninsula
itself has moved. It is as absurd as saying that a treasure is buried
under the tree where the crow is perched—when the crow may
fly away at any moment. Nor will it constitute a valid proof to
say that the conclusion is in agreement with the old continental
drift theory. What we need is a "completely independent" evidence.

This objection was put forward by the Newcastle group. What,
then, was their approach to the problem?

Locus of Polar Wandering

Their approach was to present paleomagnetic data in the following
way. From the direction of the ancient geomagnetic field preserved
in a rock from England, for instance, they calculated the ancient
position of the geomagnetic north pole. Its declination tells us the
orientation of the ancient pole, while its inclination tells us—if we
assume a dipolar field—the latitude of England relative to that
field. Therefore, we can determine the ancient position of the geomagnetic pole relative to the present position of England. The position of the ancient pole (called paleomagnetic pole) thus determined is indicated in Figure 5-10.

The Newcastle group measured, with great dispatch, the strata
in England and other European countries, and through the proce-

dure explained above, determined the position of the paleomagnetic north pole for every geologic period since the Precambrian.
The result is shown in Figure 5-11. It is interesting to see that the
more distant the period, the farther removed the pole is from its
present position. In the late Precambrian, the geomagnetic north
pole was located near the present west coast of North America and

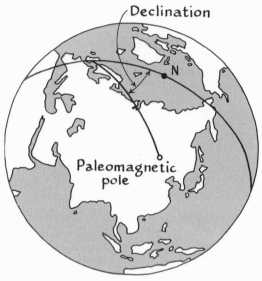

FIGURE 5–10.

Paleomagnetic pole.

then moved to the Southern Hemisphere. It was in the middle of
the Pacific in the Cambrian, passed by the northern part of Japan
in the Carboniferous and then proceeded to the present pole position. Such a curve tracing the movement of the pole is called the
locus of polar wandering.

    Thus, while the London group interpreted the discrepancy
between the past and the present geomagnetic fields as due to continental drift, the Newcastle group claimed that it was due to
polar wandering, without the continents having drifted. Recall the

locus of polar wandering mentioned in Chapter 1, the one proposed by paleoclimatologists (see Figure 1-22). A locus derived from such non-quantitative evidence as paleoclimatology today affords can hardly be very precise. Allowing for this inaccuracy, we can regard the two loci, the paleomagnetic and the paleoclimatological, as being fairly similar in broad outline. Now the one derived from

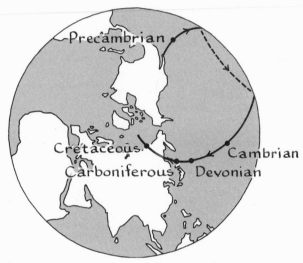

FIGURE 5–11.

Locus of polar wandering based on paleomagnetic evidence from Europe. (From figure 20 in *Continental Drift*, edited by S. K. Runcorn (1962), with permission of Academic Press Inc., New York.)

paleoclimatology is the locus of the earth's geographical pole. The one derived from paleomagnetism is the locus of the geomagnetic pole. If these two are fairly similar, does it not follow that the geographical pole and the geomagnetic pole have always coincided approximately and moved together so to speak? This concept has an important bearing on the question of the origin of the geomagnetic field.

Thus, in explanation of paleomagnetic data, the Newcastle

group invoked polar wandering instead of continental drift. Runcorn (1955) summed up his point of view as follows:

> Appreciable polar wandering seems indicated. There does not yet seem to be a need to invoke appreciable amounts of continental drift to explain the paleomagnetic results so far obtained. (*17*)

After All, There Was Once No Atlantic

Parallel to their activities in England and other European countries, Runcorn and his colleagues launched an intensive study of the North American Continent. This vast continent is also favored with strata of every geologic period. Sandstones of Arizona, shales of Texas and other rocks were measured and data accumulated quickly. In 1957, the locus of polar wandering based on American rocks was obtained. Look at Figure 5-12. According to this locus, the geomagnetic north pole was in the southeastern part of the Pacific in the Precambrian and moved westward through the Cambrian. During the Silurian period, it passed on the east side of the Philippines, moved across China in the Permian, through Northern Siberia in the Cretaceous and finally approached the present pole position. The general pattern of movement is quite similar to that of the locus obtained from Europe. The latter is shown, by way of comparison, by the dotted line in Figure 5-12. Runcorn and his colleagues looked at these two curves and wondered: allowing for the uncertainties involved in the measurement and the scattering of the data, should these two curves be considered as one, or as two roughly parallel but different curves?

Upon careful examination, they found that there is a consistent distance of about 30° in longitude between the two curves from the Precambrian to the Triassic. Is this not too systematic to be due to errors in measurement or to the scattering of the data? Our basic assumption was that there has always been only one geomagnetic north pole. What, then, does this distance signify?

At this point, the idea struck. What if the American Continent

and the European Continent moved away from each other by 30°
in longitude after the rocks had become magnetic fossils? If there
are two curves when there should be only one, the only explanation seems to be that the continents themselves have moved. What

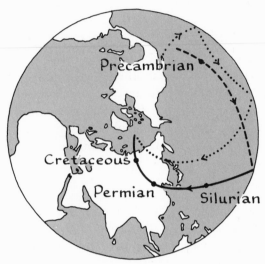

FIGURE 5-12.

Locus of polar wandering based on paleomagnetic evidence from North
America (the solid curve). The locus from Europe is shown, by way of
comparison, by the dotted curve. (From figure 20 in *Continental Drift*,
edited by S. K. Runcorn (1962), with permission of Academic Press Inc.,
New York.)

is the ancient position of the continents then? If we move the
American Continent eastward by 30° so as to make the two curves
coincide, the Atlantic almost disappears and the American Continent is joined to the European Continent. What is more, the two
curves approach one another from the Triassic to the Jurassic and
are joined by the Tertiary period. This can be explained if Europe
and America began to split apart around the Jurassic or the
Cretaceous.

Wegener had insisted that the primordial continent, Pangea, began to split apart around the Jurassic or the Cretaceous, giving rise to the present Atlantic. The old theory and the new fit together perfectly. Moreover, according to the Newcastle group, this new evidence is essentially different from the evidence on India proposed by the London group; in the latter case, paleomagnetic results can be interpreted as due to polar wandering alone, without the Indian peninsula having moved. This time, both polar wandering and continental drift are required. In order to prove the movement of two continents relative to one another, the valid approach, according to the Newcastle group, is to derive the loci of polar wandering from these two continents and find their discrepancy. Thus, the Newcastle group claimed to have obtained the first valid paleomagnetic evidence of past continental land movement.

The Background Influence

Thus, in the middle of the 1950's, the continental drift theory and the theory of polar wandering were revived by the work of English rock magnetists. This was an entirely unexpected turn of events. The old authorities who had participated in the former controversy on the continental drift and the polar wandering were not familiar with this new branch of science, paleomagnetism, while the younger generation specializing in paleomagnetism or rock magnetism had hardly heard of this dream of Wegener's which had apparently died two decades ago. What inspired the English scientists to connect paleomagnetism to the question of continental drift or polar wandering, and to devote time and energy to this work? Blackett recalls in the Introduction to *A Symposium on Continental Drift* published in 1965 that:

> . . . the survey by Arthur Holmes in *The Principles of Physical Geology* (6) certainly played a valuable part in convincing the early workers in rock magnetism that the probability that the continents had drifted was high enough to justify an intensive study of the directions of magnetizations of ancient rocks. To a physicist like myself one of the most convincing single pieces of evidence was

the Permo-Carboniferous glaciation of the southern hemisphere at about the same time that the great coal deposits of the northern hemisphere were being laid down. I remember the impression made on me by Holmes' remark that there was just not enough water in the world to produce a large enough ice-cap if the continents were then in the same relative position to each other as they are today. (*1*)

Of course, this new bombshell from paleomagnetism did not clear away all the difficulties pertaining to the continental drift theory. The opponents still objected and the neutral remained neutral. Nevertheless, interest was rekindled and the matter could no longer be ignored. We will examine some of the new objections later. Here we will only quote, to illustrate the caution that is advisable in science, the remark of Jeffreys, the great English authority on geophysics, in his fourth edition (1959) of *The Earth:*

When I last did a magnetic experiment (about 1909) we were warned against careless handling of permanent magnets and the magnetism was liable to change without much carelessness. In studying the magnetism of rocks, the specimen has to be broken off with a geologic hammer and then carried to the laboratory. It is supposed that in the process, its magnetization does not change to any important extent, and though I have often asked how this comes to be the case, I have never received any answer. (*10*)

Difficulties Caused by the Accumulation of Data

Encouraged by new clues in the mystery of continental drift, paleomagnetists extended their research to other parts of the world with a view to obtaining a locus of polar wandering from each continent. This work is now under way, and so far, results have been obtained for the two Americas, Europe, India, Australia, Africa, Antarctica, and Japan. They are plotted on the map in Figure 5-13. These loci, which trace the path of the geomagnetic north pole, ought to form one single curve if the continents have always remained fixed in their present positions. As it is, there are many curves scattered all over the globe.

So long as we had only two loci, one from North America and one from Europe, we could explain the discrepancy between them as the result of the separation of these continents. Now with so

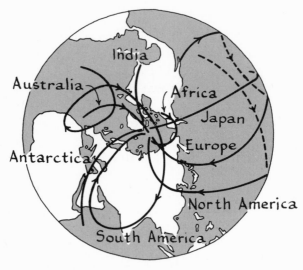

FIGURE 5–13.

Loci of polar wandering from various continents. (From figure 20 in *Continental Drift*, edited by S. K. Runcorn (1962), with permission of Academic Press Inc., New York.)

many more loci, the solution is not so easy. Which continent should be moved and how? Attempts to reconcile so many curves by mere parallel shifting of continents lead to incongruous results. Therefore, we must invoke the idea that land masses have rotated as well. (This idea was adopted in the case of the Triassic layers of England.) For instance, according to the locus obtained from Japan (see Figure 5-13), the pole was in the eastern part of the Pacific, near the Equator, in the Cretaceous period. According to the locus obtained from Europe and America, however, the pole was then already in North Siberia. If we attempt to explain this much difference by the parallel movement of Japan alone, our country ought to have been in the middle of the Atlantic in the Cretaceous. This difficulty can be easily overcome by rotating Japan clockwise.

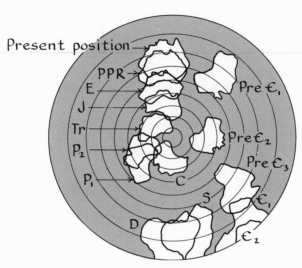

FIGURE 5–14.

Movement of Australia.

Pre Ꞓ	Precambrian	P	Permian
Ꞓ	Cambrian	Tr	Triassic
S	Silurian	J	Jurassic
D	Devonian	E	Eocene (early Tertiary)
C	Carboniferous	PPR	Pliocene (late Tertiary), Pleistocene (early Quaternary), and Recent (from the Pleistocene to the present).

(From figure 22 in *Continental Drift*, edited by S. K. Runcorn (1962), with permission of Academic Press Inc., New York.)

Another example is the movement of Australia. Look at Figure 5-14. According to this, Australia, whose latitude in the Precambrian was close to its present latitude, gradually approached Antarctica and then moved towards the Equator (Cambrian to

Devonian), reapproached the magnetic south pole (Carboniferous to Permian), and then moved across the pole to its present position. In short, it made a great circuit and came back to its original position. All the while, according to this figure, the continental mass has rotated in a complex way. The geomagnetic south pole adopted in this figure is the one derived from Europe, and with this pole fixed at the center, the movement of Australia relative to the pole was traced. The result is this elaborate movement. But is it really possible to determine so precisely the past movement of a continent through paleomagnetic method alone? The fact is that we can determine its ancient latitude but not its longitude. To make this point clear, we turn to Figure 5-15. The remanent mag-

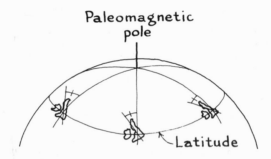

FIGURE 5-15.

netization of a rock from a given continent tells us no more than the orientation (shown in Figure 5-15) and the distance of the paleomagnetic pole relative to the continent. If we fix the paleomagnetic pole at the center of a circle, the continent can be anywhere on the circumference. In short, we can determine the ancient latitude and the orientation of the continent but not its ancient longitude. Therefore, the movement of Australia cannot really be traced without bringing in additional assumptions.

It often happens in science that while data are scarce, interpretation seems easy, but as the number of data grows, consistent argument becomes more and more difficult. When this stage is reached, we must either increase the number of data substantially or else greatly improve the reliability of each datum. However, it is one thing to increase data from ten to a hundred. It is another to increase them from a hundred to a thousand. Today, paleomagnetism seems to have reached this difficult stage.

Has Japan Been Bent?

As research went into details and met with complexities, it also brought to light as by-products some interesting information. Groups of Japanese paleomagnetists, such as N. Kawai and others at Kyoto and Osaka Universities (8), carried out extensive measurements of the remanent magnetization of Japanese rocks. Although older layers are unfit for paleomagnetic use because of violent local disturbances, Mesozoic or younger layers can still be studied. These paleomagnetists classified their numerous data into pre-Tertiary and post-Tertiary samples and plotted their magnetization direction (declination only) on the map in Figure 5-16, (A) for the post-Tertiary and (B) for the pre-Tertiary. Each arrow indicates the direction of declination and is rooted at the sampling site of the rock. In (A), the arrows all point more or less in the north-south direction, but in (B), there is a discrepancy of about 40° between the mean direction of the arrows in northeastern Japan and that of the arrows in southwestern Japan. How is this? Having seen so many interpretations of these paleomagnetic results, the reader can have no difficulty in anticipating the answer: Japan must have been bent at the middle to the extent of about 40° in the early Tertiary. This would explain the discrepancy between the two groups of arrows.

In fact, the main island of Japan looks as if it *has* bent at the middle. If we turn its northeastern half 40° clockwise, the island becomes straight. What is more, in the middle of the island there is a large geological rift called Fossa Magna ("big ditch"). The northeastern half and the southwestern half, divided by this rift, are geologically quite different.

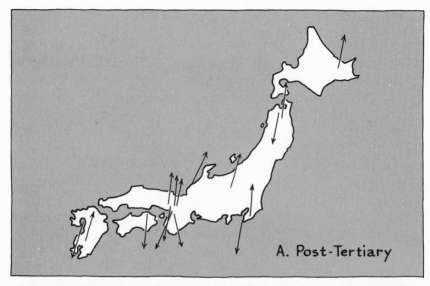

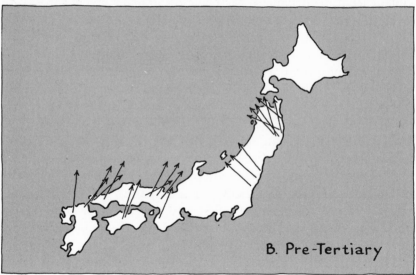

FIGURE 5–16.

Direction of remanent magnetization of pre-Tertiary and post-Tertiary rocks in Japan. (After figures 1 and 2 in Late Cretaceous crustal movements of Japan Isles estimated from the paleomagnetism, by N. Kawai and E. Abe, in *Annual Progress Report of the Rock Magnetism Research Group in Japan, 1963,* with permission of the authors.)

We must not jump to conclusions just because these facts co-incide, but this possibility is interesting in studying the geological history of Japan.

Criticism

To return to our subject, the first sweeping success of the paleomagnetic approach to the question of continental drift met with a check when different loci appeared for each continent. Some magnetists are still pushing forward, amassing more and more data, but others began to reflect that, at this rate, we might obtain a different locus not only for each continent but also for many smaller areas. The situation would become completely chaotic and hopeless. Why pursue such a course?

Certainly, it is dangerous to amass data without due consideration of their accuracy and reliability. Under such circumstances, it was but natural that some scientists began to reexamine these data (already several hundred in number) and select the reliable ones to see what could be deduced from these alone.

One such scientist, F. F. Evison of New Zealand, examined carefully the paleomagnetic data from Europe and came to the conclusion in 1961 that neither continental drift nor polar wandering is needed to account for these data (*8*). His theory was that continents expand in area by gradually flowing out from inland plateaus toward the ocean basins. The process is similar to the flow of glaciers. According to Evison, the continent expands in area by what he calls plastic flow due to gravity, and the speed of flow, greatest at the surface, diminishes with depth (see Figure 5-17). Therefore, if we assume an arrow within the crust (the broad arrow

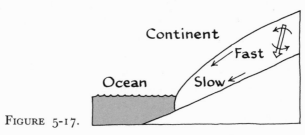

FIGURE 5-17.

in the figure), it will be gradually deflected because of this difference in the speed of flow. According to Evison, the discrepancy between the remanent magnetization direction of ancient rocks and the present geomagnetic field direction is due to this cause. This interpretation, he claimed, neatly explains the paleomagnetic data from Europe. According to this thesis, the direction of continental flow in Europe was calculated from paleomagnetic data. The result is shown in Figure 5-18. Except for one arrow which points northwestward, all the arrows point southwestward—in other words, from Northern Europe and Russia which, throughout geologic his-

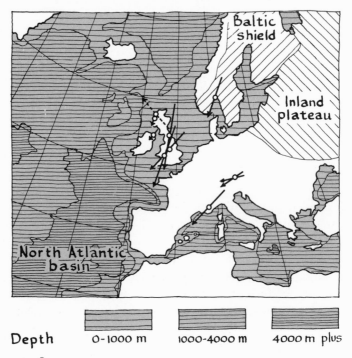

FIGURE 5–18.

Rock flow in Western Europe inferred from remanent magnetization. (After figure 2 in Rock magnetism in Western Europe as an indication of continental growth, by F. F. Evison, in *Geophy. J. vol. 4, 1961, p. 328*, with permission of the author.)

tory, have constituted the inland plateau, toward the Atlantic Basin. The speed of flow, far smaller than that of continental drift, was estimated at 10 kilometers per 10 million years.

If this interpretation is correct, the continental drift theory would not be necessary to explain the discrepancy between the past and the present geomagnetic directions. However, since the idea of continental flow under gravity is based on hypotheses without proof, this thesis has not won wide acceptance.

Another theory, proposed by F. H. Hibberd of Australia in 1962, attempts to reconcile all the loci from various continents (8). The one finally obtained, through statistical procedure, is shown in Figure 5-19. Hibberd held that all the discrepancies between the actual loci and this one result from secondary magnetization (see page 159). This is in fact a challenge to the very foundation of

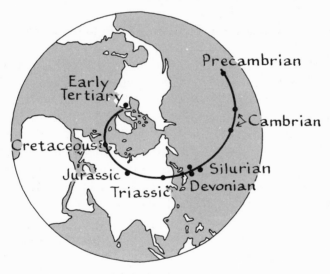

FIGURE 5–19.

Locus of the paleomagnetic pole in the present Northern Hemisphere according to Hibberd. (After figure 5 in An analysis of the positions of the earth's magnetic pole in the geological past, by F. H. Hibberd, in *Geophys. J.*, 6, No. 2, *1962*, *p. 230*, with permission of the author.)

paleomagnetism. Hibberd assumed that the magnetic north pole has moved along the locus in Figure 5-19 while none of the continents has moved—that is, there has been polar wandering but not continental drift. The rocks must have acquired natural remanent magnetization in the direction of the geomagnetic field at the time of their formation. Suppose they acquired subsequently, over and above this primary magnetization, secondary magnetization in the direction of the present earth's field, would all the discrepancies be explained as the result of this dual magnetization? Hibberd investigated this, and, surprisingly enough, the result turned out to favor his hypothesis: in the majority of cases, the discrepancies could be explained in this way. Hibberd's argument is as yet unrefined in that it ignores all the differences in the intensity of secondary magnetization depending on types of rocks. Nevertheless, we cannot flatly reject this argument; the paleomagnetic data, hitherto adopted somewhat indiscriminately, may well include those which are in fact unusable because of this secondary magnetization.

In 1963, there appeared another criticism of the validity of polar and continental movement hypotheses. American scientists, J. W. Northrop and A. A. Meyerhoff (8) argued that if the earth's field has not always approximated a dipolar field, there need be neither continental drift nor polar wandering. Indeed, it is only as far back as the late Mesozoic that the earth's field is proved to have been a dipolar one (see page 142), and it is around this period or earlier that the loci of polar wandering from various continents begin to split apart. Therefore, according to these scientists, it would be more reasonable to suppose that the earth's field had been non-dipolar and had had multiple magnetic poles prior to the Mesozoic. The idea of a non-dipolar field has always existed in paleomagnetism and that is why paleomagnetists always qualified their statements by saying "if we assume a dipolar field." When data are collected to indicate the possibility that there existed contemporaneously a number of magnetic poles, would it not be more natural to accept the concept of multi-polar field rather than to try to reconcile so many places of poles? According to the proponents of this idea, the geomagnetic field such as the one postulated by the dynamo theory, began to be formed towards the end of the Precambrian, underwent complex processes, and took its present form only towards the late Mesozoic. So far, this hypothesis of non-dipolar field cannot be refuted by paleomagnetic data alone.

Blackett's Group: a Reexamination of Premises

Thus, in the early 1960's, the tendency was to reexamine paleomagnetic data and principles. One typical example lies in the work of Blackett and his colleagues at London University.

This group had kept silent when, in 1957, the Newcastle group had objected to their interpretation of paleomagnetic data. In 1960, they spoke up again. As we know, the London group, in interpreting paleomagnetic results, had always confined their argument to the rotation and the latitudinal movement of land masses. For this, they said, there had been an excellent reason. The paleomagnetic data can determine the ancient latitude of a continent and its ancient orientation relative to the geomagnetic pole, but not its ancient longitude. Therefore, from the two loci of polar wandering obtained by Runcorn, it is logically impossible to deduce the separation of the North American and the European Continents in the east-west or longitudinal direction. Indeed, we can make the two loci coincide by latitudinal and rotational movements of the continents alone. Therefore, according to Blackett, Runcorn's conclusion that Europe and North America had moved apart, giving rise to the Atlantic, was not based after all on any "independent evidence," but was selected arbitrarily from a variety of possible conclusions so as to fit Wegener's theory.

Blackett and his colleagues reexamined in their own way what can really be deduced from the paleomagnetic data (8). Look at Figure 5-20. In this figure, all that we know directly from observed data—the ancient latitudes and the orientations of continents at various geological periods—are given for Europe, North America, India and Australia. For convenience, a town located approximately at the center of the continent was selected as a reference point, Paris for Europe, Denver for North America, Nagpur for India, and Alice Springs for Australia, and the ancient latitude and the orientation of each continent are given in terms of these values for the reference town. Ancient latitudes are plotted

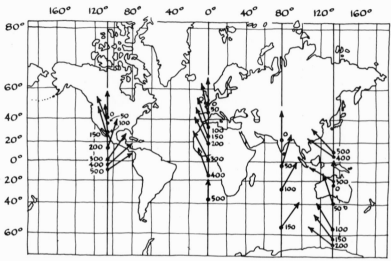

FIGURE 5–20.

Ancient latitudes and orientations of Europe, North America, India, and Australia at various geological periods. The number against each point indicates its age in millions of years. (After figure 8 in An analysis of rock magnetic data, by P. M. S. Blackett, J. A. Clegg, and P. H. S. Stubbs, in *Proc. Roy. Soc. A. 256, 1960, p. 312,* with permission of the authors.)

in terms of the present geographical latitude—in other words, without considering polar wandering. This method of presentation is just the opposite of that in Figure 5-13 which does not take into account any movement of continents. No ancient longitudes are given because they are indeterminable. This does not of course mean that there has been no longitudinal change. Figure 5-20 contains all the information provided by paleomagnetic data.

It is clear from this figure that all the continents have drifted northward throughout geologic history. The speed of drift is about

0.2° ∼ 0.8° in latitude or about 20 ∼ 90 km per million years. Particularly for the last 300 million years, Europe and North America have moved northward abreast. They have rotated, however, in opposite directions. Europe has rotated about 50° clockwise relative to North America. The movement of Australia is somewhat complex. India has moved the farthest of all.

According to Blackett and his colleagues, paleomagnetic data should be presented in this way to indicate the continental movement and rotation. Certainly, it is impossible to explain all these facts by polar wandering alone. Therefore, what we can say with certainty at the present stage is this: unless we bring in the hypothesis of non-dipolar field, some kind of continental movement is certainly needed to explain paleomagnetic data. However, as to the question of the direction in which the continents have moved, rock magnetism can provide no information about longitudinal movement.

There are also problems concerning the reliability of the paleomagnetic method. For instance, the direction of remanent magnetization can sometimes be affected by various stresses acting on naturally occurring rocks. Sometimes, in sedimentary or metamorphic rocks, the ferromagnetic mineral grains or their crystal orientations are aligned in a certain direction. In such a case, the direction of remanent magnetization is influenced not only by the external geomagnetic field direction but also by the direction in which the grains or the crystals are aligned. Such a rock would not be a faithful fossil of the past geomagnetic field.

Of course, rock magnetists are doing their best to overcome these difficulties. After its brilliant start, rock magnetism is now preparing for a higher stage of development.

Meanwhile, we must next consider the problem of drift from another angle.

6

Is the Earth Heating or Cooling?

The Earth as a Heat Engine

We have given much time to consideration of problems related to continental drift. The ideas of continental drift and polar wandering may be true, or they may turn out to be illusions. Numerous evidences have been advanced and rejected. As many hypotheses have been proposed, then rebutted.

Our theme hinges upon the fundamental question of earth science, the history of the earth. The course of that history seems to have been governed by the thermal energy of the earth, the topic of this chapter, for study of the earth's thermal history may give us some idea as to whether continental drift is at all a likely event or an utter improbability.

The most important energy source of the earth is thermal energy such as is released by radioactive disintegration of matter. Of course, energy takes other forms: gravitational energy, the kinetic energy of rotation and revolution and so on. Eventually, however, such energy is transformed into heat in the earth's interior and warms up the earth. Heat is gradually transferred from the earth's interior to the surface, and escapes into space. In the course of such outflow, thermal energy sets off various geological phenomena. In this sense, the earth is a "heat engine."

The reader will recall the theory of mantle convection, proposed as the mechanism of continental drift (Chapter 2). If there is convection in the mantle, it may well cause the displacement of continents. Displacement of huge land masses would mean a considerable change in the distribution of the mass of the earth, which in turn would disturb the equilibrium of the earth's rotation. This may possibly result in polar displacement. But is there any evidence, from the study of the earth's thermal state, that convection occurs in the mantle?

We have from time to time touched on the thermal aspect of the earth. This chapter will present a more systematic analysis of the history and the present state of the earth's thermal energy.

To start with, is the earth heating or cooling?

Heat from the Ground

The simplest way of studying the earth's thermal state is to drill a hole to measure internal temperature.

As we know from daily experience, the temperature at the earth's surface fluctuates constantly; it is high during the day and low at night; it varies with the season. These meteorological temperature changes mainly depend on the heat from the sun, and their effect becomes negligibly small at the depth of 30 meters. We are concerned with the earth's temperature at greater depths, so its measurement must be made at as great a depth as is practically possible. Mine shafts, tunnels, and oil wells have been utilized for this purpose. Such measurements show that the earth's temperature at any locality increases with depth. This is true not only of volcanic regions but also of non-volcanic areas.

The rate at which the earth's temperature increases with depth is called *geothermal gradient*. In non-volcanic regions, the geothermal gradient is, on the average, 3°C per 100 meters of depth. This is a rough average, for actually the gradient is highly variable from place to place, ranging from less than 1°C to more than 5°C per 100 meters.

That is about all the information we can obtain from direct measurement.

If we assume that the thermal gradient continues at the same rate to greater depths, the temperature at the earth's center would be about 200,000°C. Such calculation is of little value, however, for as we shall see (page 201), there is a sounder method of estimating the earth's internal temperature.

When there is a temperature difference within a substance, heat flows from the hotter to the cooler region. The greater the temperature difference (in other words, the thermal gradient), the greater the amount of heat flow. The amount of heat flow also depends on the thermal conductivity (see page 83) of the substance. In short, the quantity of heat flow through a substance is the product of the thermal gradient and the thermal conductivity of the substance. Therefore, if we know the thermal conductivity of the strata in which the thermal gradient was measured, we can calculate the quantity of heat flowing through the strata. The quantity of heat that flows from the earth's interior to its surface and escapes into space is called *terrestrial heat flow*. This is the quantity of thermal energy that the earth is losing, in other words, the expenditure of the earth's heat budget.

English geoscientists such as Bullard first recognized the importance of terrestrial heat flow and started its measurement in the early 1930's. Thus Bullard is a pioneer in geothermy as well as an authority on geomagnetism.

Data on terrestrial heat flow are still scarce; by 1960, fewer than one hundred measurements had been reported. These data, however, indicated a significant phenomenon: for all continental regions of the world, the terrestrial heat flow is approximately the same, about 1.5×10^{-6} cal/cm² sec. Although the thermal gradient varies considerably from place to place, the thermal conductivity is low where the thermal gradient is high so that the product of the two, the terrestrial heat flow, is more or less constant.

This heat in continental regions is mostly generated in the earth's crust, and not in the mantle, as we explained in Chapter 2 (see page 82). To review briefly, radioactive elements which generate heat are most abundant in the granitic rock, less abundant in the basaltic and least abundant in peridotite (see Table 2-2, page 81). Assuming that the continental crust consists of 15 km of granite resting on 15 km of basalt, the amount of heat generated in the part of the crust underlying a surface area of 1 cm² would be approximately 1×10^{-6} cal/sec. The terrestrial heat flow on land is 1.5×10^{-6} cal/cm² sec. It follows then that most of the heat which escapes from the land surface of the earth is generated in the continental crust.

Heat from the Ocean Floor

What about terrestrial heat flow at sea which covers more than two-thirds of the earth's surface? The measurement of the thermal gradient and the thermal conductivity of the materials of the ocean floor, beneath several thousand meters of water, poses practical difficulties. There is one advantage, however; below the depth of 2000 meters, the water temperature is constant—about 1°C—all the year round. Therefore, there is no need to drill a deep hole for the sake of eliminating the effects of daily and seasonal temperature changes; any hole from 50 cm to 1 m deep would do.

In the early 1950's, Bullard and his colleagues succeeded in devising an instrument for the measurement of terrestrial heat flow at sea (9). Look at Figure 6-1. A steel probe a few meters long (A in the figure) contains, at points 1 and 2, thermal sensing elements (thermocouples or thermistors). The temperature difference between the points 1 and 2 is measured and automatically recorded by the recording instrument in the steel case B. The whole instrument is lowered overboard on a steel cable. The probe easily penetrates the ocean floor because its surface is generally covered with soft sediments. At the same time, these sediments are collected by a coring instrument for the measurement of their thermal conductivity. With this instrument, Bullard and his colleagues measured the terrestrial heat flow in the Pacific and the Atlantic oceans.

Prior to the measurement, the heat flow through the ocean floor was expected to be far smaller than that through the continents, for the good reason that, beneath the oceans, there is only a thin crust, at most a few kilometers thick. This crust, moreover, consists of basaltic rocks which are less radioactive than granite. Immediately under this crust is the mantle (see Figure 1-13, page 39). With such a structure, the radioactive heat generation in the

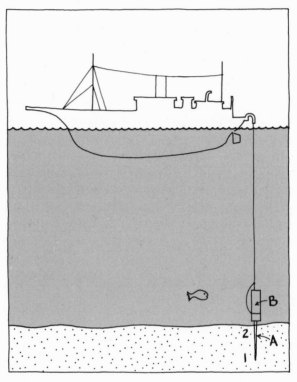

FIGURE 6–1.

Schematic drawing of the instrument for the measurement of heat flow at sea.

ocean floor must be several times less than that in the continental region. Hence, the heat flow through the ocean floor must be that much lower than the heat flow through the continental regions.

Contrary to this "reasonable" expectation, actual measurement at several sites in the Pacific and the Atlantic revealed that the heat flow through the ocean floor is roughly equal (about 1.5×10^{-6} cal/cm² sec) to that through the continental regions. What then of the difference in the crustal structure and hence in the quantity of the heat supply? Which of the foregoing assumptions had been wrong?

Thus, another great riddle appeared in earth science. In the authors' opinion, this was one of the most important events in the geophysics of the post-war years. The riddle is not only important in its own right; it attracted general attention to the thermal aspect of the earth. The problem, involving the question of the radioactive distribution within the earth, impinges upon geochemistry. Moreover, it stimulated the study of the ocean floor (see next chapter) which has recently made great strides forward.

Thermal Convection in the Mantle

Figure 6-2 illustrates our problem. On land, heat is generated in the thick crust. At sea, there is no such heat source in the crust.

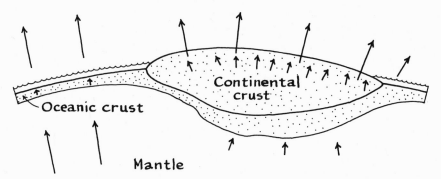

FIGURE 6-2.

We must infer then that the mantle under the ocean contains a considerable amount of radioactivity and generates heat equivalent in quantity to the heat produced in the continental crust. This is a simple inference, but its implications are far-reaching. For one thing, it would imply that the chemical composition of the mantle is different under land and sea. Seismologists testify that, below the Mohorovičič discontinuity (see page 37), earthquake waves travel at about the same speed, whether under continents or under oceans. Therefore, it has been a well-established fact in earth science that the mantle is of uniform composition all over the globe. The result of thermal measurement, however, suggests that the composition of the mantle differs considerably from one locality to another. This idea seems reasonable enough, for is it not unnatural that the great surface difference between land and sea disappears at the depth of only 30 km?

Let us move on to the second point. A rock is essentially a poor heat conductor. Its thermal conductivity is no more than a hundredth of that of metal. Therefore, even if the mantle contained a sufficient amount of radioactive elements, the heat generated therein would not easily come up through the mantle (which is after all 2900 km thick) to the earth's surface. The process of conduction is so slow that heat generated at the depth of a few hundred kilometers at the time of the earth's birth (4.5 billion years ago) is only today reaching the earth's surface. Heat generated at greater depths does not reach the surface, but warms up the mantle. A simple calculation revealed that if the amount of radioactivity great enough to account for the terrestrial heat flow at sea is distributed beyond the depth of a few hundred kilometers, the lower portion of the mantle would have been melted by this time. This is contradictory to the undeniable seismic evidence that the mantle is a solid. In order not to melt the mantle, the radioactive heat source must be concentrated in the top few-hundred kilometers of the mantle.

In that case, however, the upper layer of the mantle would have to be many times more radioactive than ordinary peridotite. It is rather unlikely that any peridotite contains that much of radioactive elements.

The difficulty can be solved if the radioactive elements are widely distributed over the mantle and at the same time there exists in the mantle some mechanism by which heat is transferred more effectively than by mere conduction. Then the mantle would neither melt nor have to contain this inordinate amount of radio-activity concentrated in the upper layer.

There are three mechanisms of heat transfer, thermal conduction, thermal radiation and thermal convection. As we saw in Chapter 3, atoms that compose all matter make thermal motions because they have thermal energy. When heat is transferred through a propagation of such atomic thermal motion, the process is called thermal conduction. Atoms also radiate energy in the form of electromagnetic waves. This phenomenon is called radiation. Heat from the sun, for instance, comes to the earth by radiation. The third mechanism, thermal convection, was explained in Chapter 2. In this case, heat is transferred by the motion of fluid substance (see page 85).

Geophysicists tell us that at a high temperature, such as that prevailing in the mantle, heat transfer by radiation is dominant over conduction. Therefore it is possible that, by the mechanism of radiation, heat is transferred more effectively in the mantle than by mere conduction.

Another solution, suggested by Bullard and his colleagues, is that the heat in the mantle is transferred by convection. We saw in Chapter 2 that, on the vast geologic time scale, thermal convection can occur in a solid body like the mantle and that it disposes of heat effectively. Thus, the equivalence of the terrestrial heat flow through continents and the ocean floor suggested the possibility that there may be thermal convection in the suboceanic mantle. The hypothesis of mantle convection was thus given a corroborative support from the measurement of terrestrial heat flow. Further development will be discussed in the next chapter.

Temperature in the Crust and the Mantle

A bore-hole used for the measurement of the earth's temperature is at most a few kilometers deep. How can we estimate the earth's temperature beyond that depth?

For the temperature in the crust, the observed value of the surface heat flow gives a clue. If the thermal gradient is 3°C per hundred meters all the way to the bottom of the crust (30 km deep), the temperature there would be 900°C (the surface temperature is assumed to be 0°C). As the heat escaping from the surface is mostly generated in the crust (at least in continental regions), there is little heat flow from the mantle to the crust. Therefore, the thermal gradient at the bottom of the crust must be smaller than that at the earth's surface. Similarly, the thermal gradient at any given depth of the crust must be smaller than the surface gradient, the difference being due to the radioactive heat produced above that depth. As gradient at any depth of the crust is thus smaller than the surface gradient, the actual temperature at the bottom of the crust must be lower than 900°C.

On the other hand, experiments have shown that the thermal conductivity of rocks tends to decrease as temperature rises. Therefore, the greater the depth, the greater the thermal gradient required to convey the same quantity of heat. This fact offsets to some extent what we mentioned just above, the decrease of thermal gradient with depth. Taking both factors into consideration, the temperature at the base of the crust (at the depth of 30 km) is estimated at 600° ∼ 800°C. At sea, since the crust is only about 5 km thick, the temperature at its base would be 150° ∼ 200°C. As the value of terrestrial heat flow varies considerably from place to place, particularly in oceanic regions (see page 227), the crustal temperature would also vary with locality.

What about the temperature in the mantle? As the mantle is 2900 km thick, its internal temperature cannot be estimated from surface measurement. We must resort to some indirect method.

The earth is not heated by radioactive disintegration alone. Generally, the temperature of a substance is raised by compression. A diesel engine, for instance, utilizes the fact that the gas in the cylinder, when it is suddenly compressed, reaches the ignition point and explodes. This temperature increase resulting from compression, i.e., without any external supply of heat, is called *temperature increase due to adiabatic compression*. Since the pressure in the earth is tremendous (about 4000 kb at its center), there would certainly be temperature increase resulting from this adiabatic compression. If we can estimate this temperature increase, we can set the lower limit for the earth's temperature, for the earth would heat up to this level by compression alone without the aid of radioactive heat.

The temperature increase that is due to adiabatic compression at a given pressure is determined by the thermal expansion coefficient (the rate at which a substance expands with the rise in temperature) and by the "specific heat" (heat required to raise the temperature of a substance by 1 °C). The thermal expansion coefficient and the specific heat are intrinsic properties of a substance and vary with pressure and temperature. How can we estimate the thermal expansion coefficient and the specific heat of the mantle substance?

The best known property of the mantle substance is the speed at which it transmits seismic waves. Using principles and formulae of solid state physics, J. Verhoogen (20) and others have attempted to estimate the thermal expansion coefficient and the specific heat of the mantle substance from the speed of seismic waves. According to their results, the temperature at the mantle-core boundary would be 1000° $\sim$ 1200°C by adiabatic compression alone. This, then, is the lower limit of the mantle temperature.

As for the upper limit, the solidity of the mantle gives a clue, for the temperature therein must be lower than the melting point of the mantle substance. The most promising candidate for the mantle substance is peridotite. Its main constituent mineral is olivine (Mg_2SiO_4). Olivine, when heated in the laboratory, melts at about 1900°C under normal pressure. The melting point generally increases with pressure. Experiments have shown that the melting point of olivine is raised by 4.7°C per 1 kb. At this rate, its melting point would be about 1950°C at the base of the crust.

This rate of increase, however, does not apply to higher pressures. With our present experimental technique, the highest pressure that can be produced in a laboratory is about 100 kb which, in the earth's interior, is the pressure at the depth of only 300 km. For the estimation of the melting-point at greater depths, we must seek the aid of seismic data and solid state physics.

Unfortunately, the phenomenon of melting is one of the least well-studied domains of physics. Among the few existing theories on melting is the one proposed by F. A. Lindemann around 1910. The atoms of a solid vibrate because they have thermal energy. As the thermal energy increases with temperature, the amplitude of vibration increases till the neighboring atoms collide with one another. This breaks down the crystal structure of the substance; in other words, the substance melts. The frequency of the vibration of atoms at this point is called *critical frequency*. According to Lindemann, the melting point of a substance is proportional to the square of its critical frequency.

R. Uffen (*20*) showed that the critical frequency of a substance can be estimated from the speed of P and S waves travelling through the substance. Through this method, he estimated the melting point of the mantle (near its boundary with the core) at 5000°C. This can be taken as the upper limit of the temperature at the base of the mantle.

Apart from the upper and the lower limits, there have been attempts (H. P. Coster, 1948) to estimate the actual temperature of the mantle from the distribution of electrical conductivity of the mantle substance (*20*). Peridotite is an extremely poor electric conductor under normal temperature. However, it is a so-called semi-conductor, and its electrical conductivity increases with temperature. Therefore, the distribution of electrical conductivity within the mantle may provide a clue to the distribution of temperature.

The distribution of electrical conductivity can be determined from the rapid variations in the geomagnetic field, such as we see in daily variation or magnetic storms (see page 102). These, as we saw, are due to external causes.

In general, a change in the external magnetic field induces electric currents in an electric conductor, which, in turn, produce a magnetic field. This phenomenon, called electromagnetic induction, is one of the fundamental principles of electromagnetism. Therefore, if the mantle substance has some electrical conductivity, any change in the external magnetic field would induce an internal magnetic field in the mantle. Hence, the magnetic fluctuations observed on the earth would be a combination of the variation of the external magnetic field and that of the internal field.

It is possible to distinguish the external magnetic change from the internal one, by a mathematical analysis of the geomagnetic data from all over the world. Once we know how much external change induces internal change, we can, through the application of electromagnetic principles, estimate the electrical conductivity within the mantle. Fortunately, short-period fluctuations yield information about the electrical conductivity of the shallower part and long-period fluctuations about the deeper part. Therefore, by studying magnetic fluctuations of various periods, we can find the variation of electrical conductivity with depth and hence the variation of temperature with depth. For instance, fluctuations with a period of one hour give a clue to the electrical conductivity at the depth of 200 ∼ 300 km, while daily fluctuations with the period of 24 hours yield information about the depth of 1000 km. For information about the bottom of the mantle, a fluctuation with a period of ten years is required. For this reason, the temperature distribution has been determined only down to the depth of 1500 kilometers. At this depth, according to Rikitake (15, 20), the temperature of the mantle is 2500° ∼ 2590°C.

It is interesting to note how thermal, electric, and magnetic phenomena are interlinked; magnetic fluctuations help to determine the distribution of electrical conductivity which, in turn, helps to determine temperature distribution.

Temperature in the Core

Delving still further, we reach the core. The outer core is in a liquid state, and its chief component is probably iron (see page 38). Therefore, the melting point of iron under pressure prevailing in the core would be the minimum temperature of the outer core. Seismological data suggest that the deeper part of the core—the inner core—is solid. If this is true, the melting point of iron would give the maximum temperature for the inner core. In short, once the melting point of iron is determined, the temperature in the liquid part would be higher, and that in the solid part lower than this figure; therefore, we can assume that the actual temperature of the core is somewhere around this melting point.

What, then, is the melting point of iron under pressure prevailing in the core? As pressures of 1000 kb cannot as yet be produced in a laboratory, the melting point is estimated from Simon's formula. This is an empirical formula on the variation of the melting point of metal with pressure, developed by F. E. Simon. It is empirical in the sense that it was not derived purely by theory, but determined so as to explain experimental results satisfactorily. Therefore, in applying Simon's formula to the pressure prevailing in the core, we are assuming that the formula based on experiments of 100 kb is valid at a pressure one order of magnitude higher, 1000 kb. Although this is not a sound procedure, it is, to date, the best method available for estimating the melting point of iron under high pressures. Moreover, there is a justification for using this method. Sodium is more compressible than iron; the volume change of sodium under the pressure of 100 kb is as great as or greater than the volume change of iron under the pressure of 1000 kb. Since Simon's formula has been experimentally verified for sodium under the pressure of 100 kb, it is not unreasonable to suppose that the formula is also valid at a pressure that would produce comparable volume change in iron, that is, 1000 kb.

Through this formula, Simon obtained the value of 3700°C for the melting point of iron at the earth's center. Estimates obtained by others range from 3000°C to 6000°C. (See page 203.)

Earth's Temperature: a Summary

Estimates of the earth's temperature (20) obtained by various methods are summarized in Figure 6-3. Temperature is taken along the ordinate (the vertical axis) and the depth along the abscissa (the horizontal axis) in this figure. The crust, being only 30 km thick, appears as a thin strip in the left-hand corner. In the mantle, curve U (which stands for Uffen) indicates the variation of the melting point of the mantle substance with depth. The actual temperature in the mantle must be below this curve. On the other hand, it must be above curve S (Shimazu) which indicates an example of the temperature due to adiabatic compression alone.

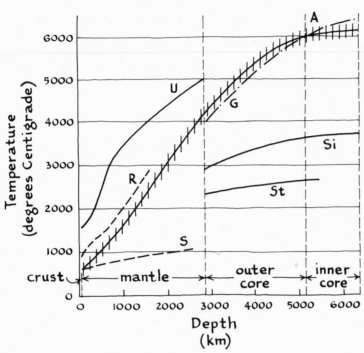

FIGURE 6–3.

Estimated temperature in the earth's interior.

Curve R (Rikitake) which shows an example of the actual temperature estimated from electrical conductivity, does indeed fall between curves U and S.

In the core, curves Si, St, and G indicate the estimated values of the melting point of iron obtained by Simon, H. M. Strong, and J. J. Gilvarry respectively. We will adopt curve G, based on the most recent data (1957), as the best approximation. Temperature at point A, the boundary of the liquid outer core and the solid inner core, must be equal to the melting point of iron under pressure prevailing there. This temperature is about 6000°C.

At the boundary of the core and the mantle, the actual temperature must fall somewhere between curves G and U, perhaps around 4200°C. By joining those estimated values, Gilvarry obtained the hatched curve. This curve was given a certain breadth because these values, based on estimates, are still uncertain. Probably there is greater uncertainty involved than is shown here, for, of all the physical properties of the earth, temperature is one of the least well-known. It seems, however, quite certain that the temperature at the earth's center is below 10,000°C.

Origin of the Earth

Having considered the present thermal state of the earth—how much heat it is losing and how its internal temperature varies with

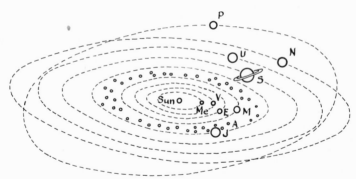

FIGURE 6-4.

Schematic drawing of the solar system.

Me	Mercury	J	Jupiter
V	Venus	S	Saturn
E	Earth	U	Uranus
M	Mars	N	Neptune
A	Asteroids	P	Pluto

depth—we now turn back to the first page of the earth's history, the birth of the earth.

As we know, the earth is one of the ten planets revolving around the sun (Figure 6-4). The planets of the solar system differ considerably in size and periods of rotation; for instance, Mercury, the innermost planet, has a small diameter (0.38 times that of the earth), but a long period of rotation (88 days); Jupiter, on the other hand, whose distance from the sun is five times that of the earth, has a diameter 11 times larger than the earth, and rotates with a period of less than ten hours. Despite these differences, these planets have many common features; they all revolve in the same direction and almost on the same orbital plane; in most cases, the direction of rotation agrees with the direction of revolution.

These facts suggest that the solar system was born as a unit. If so, the earth's origin is inseparable from that greater mystery, the origin of the solar system. This is indeed a most intriguing problem of astronomy and we can hardly do justice to it in a short section of this book. We will only outline two main schools of thought to see what light they will throw on the earth's history.

A Ball of Hot Gas?

One school of thought postulates that hot gaseous matter which once formed a part of the sun broke away for some reason, and cooled and solidified to form the planets. This is the "hot-origin" hypothesis.

Because the earth is hotter in its interior than at the surface, it was vaguely imagined that the whole earth had been hotter at its birth and has subsequently cooled. Apart from this notion, the main supporting evidence of the hot-origin hypothesis was that it could explain why, in most planets, the direction of rotation agrees with that of revolution.

Opinions are divided, however, as to how the hot gaseous matter broke away from the sun. Numerous hypotheses were proposed from the turn of the present century to the 1930's, but all assumed that some cataclysmic event had taken place; for instance some great celestial body might have approached the sun and exerted an immense force of attraction, as a result of which a part of the gaseous matter was drawn out; perhaps the celestial body had actually collided with the sun. Indeed some cataclysmic event would have been required to draw out from the sun—against its immense force of gravity—a large mass.

What would happen to the earth subsequent to such a birth? The hot gaseous matter would radiate heat from its surface and cool rapidly. Within the gaseous body, thermal convection would be set up and convey heat effectively to the surface. According to one calculation, such gas body would cool and liquefy in fifty or sixty thousand years, a mere moment in the earth's life span. The liquid earth would cool further because of convection and the thermal radiation from the surface. Liquid substances move about freely, so, at this stage, lighter material would have risen to the surface and the heavier material settled to the center. The layering of the earth—into core, mantle, and crust—would have taken place at this early stage.

When the temperature fell to the melting point of the terrestrial substance, solidification began. Figure 6-5 illustrates the process of solidification suggested by J. A. Jacobs of Canada (9). The solid curve shows the variation of the melting point of the terrestrial substance with depth. The curve is discontinuous at the mantle-core boundary, because the mantle substance and the core substance have different melting points. The dotted curves show the actual temperatures of the earth at various stages of cooling. The temperature is higher in the core than in the mantle because of adiabatic compression (see page 198).

As the earth cooled, solidification would have started at the point where the dotted curve first intersects the solid curve. Thus, according to Jacobs, solidification began at the center of the earth. A solid inner core continued to grow until the dotted curve intersected the solid curve twice, once at A, the mantle-core boundary, and again at B, as shown in Figure 6-5. As the earth cooled further, the mantle began to solidify from the bottom upward. The liquid layer between A and B was trapped, insulated above by the solidifying mantle and below by the already solid inner core, and had little chance of cooling. That is presumably why the outer core has remained liquid to this day.

Thus, our earth acquired a solid inner core, a liquid outer core and a solid mantle. Its subsequent history is one of slow but

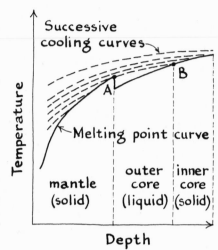

FIGURE 6–5.

(From figure 5–3 in *Physics and Geology*, by Jacobs, Russell, and Wilson (1959), with permission of McGraw-Hill Co., New York.)

progressive cooling. Such is the outline of the earth's thermal history, according to the hot-origin hypothesis.

The hotter the earth, the more lively the geologic activities. Therefore, orogeneses, earthquakes, and volcanic activities must have been more lively in the earlier part of the earth's history. This picture of the earth's history was the basis for one of the chief objections to Wegener's theory, namely, that a great event like the continental displacement could not have occurred in so recent a period as the Mesozoic (see page 69). According to the hot-origin hypothesis, a great event like the disposition of continents and oceans ought to have taken place at an early stage of the earth's history.

During the 1940's, however, the tide turned against the hot-origin hypothesis. Astronomers, such as L. Spitzer, theoretically studied whether hot gas, spurting from the sun into space, would really solidify and develop into planets; they proved that such gas would dissipate before it could possibly cool and solidify into planets.

Another weakness of the hot-origin hypothesis was that it required a cataclysmic event such as collision between celestial bodies. There are a formidable number of celestial bodies in the universe, but the vastness of the universe is still more formidable. If we compare the sun to a poppy seed, the nearest celestial body, another poppy seed, would be 30 km away, and each would be moving 50 ~ 60 cm annually. Such relatively distant bodies would rarely, if ever, collide with one another. According to one calculation, collision between any two celestial bodies could not have occurred more than a few times in the whole universe since its inception. Is it reasonable to ascribe the origin of the solar system to such rare chance? Is there not a more natural explanation?

An Aggregate of Dust?

The second hypothesis, the dust-origin hypothesis, claims to provide a more natural explanation. Interstellar space is very nearly a perfect vacuum, far more so than any vacuum we can produce in a laboratory. Even then, there are about one hundred atoms moving in a liter of cosmic space. Besides, there are fragments of matter called cosmic dust floating in the universe. According to the dust-origin hypothesis, the primordial sun collected—by its force of gravity—a great quantity of these dust particles into a fairly dense cloud around itself. The particles would collide with one another and aggregate because of their viscosity. The larger particles would collect smaller ones by their force of gravity and thus grow into larger masses. This, according to the dust-origin hypothesis, is how the planets were formed.

Actually, a hypothesis similar to this one, called the Kant-Laplace nebular hypothesis, had existed prior to the hot-origin hypothesis. The nebular hypothesis, however, had to yield to the latter, because in the days of Kant and Laplace the nebular hypothesis could not satisfactorily explain how the planets came to revolve at high speeds around the sun. In this respect, too, the proponents of the new dust-origin hypothesis took pains to provide a satisfactory explanation. The point of primary interest to us, however, is that, according to the aggregate hypothesis, the earth was formed at a relatively low temperature. What then would be the subsequent thermal history of the earth?

A Meteoritic Earth?

To begin with, what substances composed the earth at its birth? This composition determines the quantity of radioactive heat supply and hence has important bearing on the earth's thermal history. Unfortunately, the chemical composition of the earth—the kinds of elements that compose it and their relative abundance— cannot be determined directly, for only surface rocks are available for chemical analyses.

For reasons which will soon be given, many scientists now consider that the earth's original substance was similar to a kind of meteorite called chondrite.

A meteor is a piece of cosmic matter that enters the earth's atmosphere. It is well-known that the shooting light of the meteor across the night sky is due to the heat caused by its friction with the air. Occasionally, a meteor succeeds in reaching the earth's surface, and such a body is called a meteorite. Meteorites can be classified roughly into two kinds: iron meteorite composed chiefly of iron and nickel; and stony meteorite composed of silicate minerals. Chondrite is a typical stony meteorite. Table 6-1 shows the average chemical composition of chondrite.

It is thought, for the following reasons, that chondrite represents the chemical composition of the earth. For one thing, it is possible that meteorites are fragments of a former planet which, for some reason, was disrupted and dispersed. The orbital movement of meteorites suggests that they are not wandering at random in infinite space but are members of the solar system.

Secondly, terrestrial material and chondrite are quite similar

TABLE 6-1 *

RELATIVE COMPOSITION BY WEIGHT
OF CHONDRITIC METEORITE

Silicon (Si).......................... 1.00×10^6
Magnesium (Mg)................... 9.34×10^5
Iron (Fe)........................... 7.12×10^5
Sulfur (S).......................... 1.04×10^5
Aluminum (Al)..................... 7.91×10^4
Calcium (Ca)....................... 5.52×10^4
Sodium (Na)....................... 4.94×10^4
Nickel (Ni)......................... 3.64×10^4
Chromium (Cr)..................... 7.70×10^3
Potassium (K)...................... 5.94×10^3
Manganese (Mn)................... 5.64×10^3
Phosphorus (P)..................... 4.55×10^3
Titanium (Ti)...................... 2.27×10^3
Cobalt (Co)........................ 2.21×10^3

* From Table 5 in *Researches in Geochemistry*, edited by P. H. Abelson (1959), with permission of John Wiley & Sons, New York.

in isotopic composition (isotopes, you will recall, are atoms with the same atomic number but with different mass numbers). *Chemical* composition refers to the proportion of different elements within a substance, a certain percent of oxygen, a certain percent of iron, etc. *Isotopic* composition refers to the proportion of isotopes within a single element (for instance, the proportion of Fe^{56} and Fe^{54} in iron). The chemical composition of the terrestrial rock may have changed from that of the initial terrestrial substance by various chemical changes, but the isotopic composition of individual elements must have remained relatively unchanged. If identical twins are reared apart, one may grow richer and the other poorer. Still, there would be many indications of their common parental origin in their identical blood-type and other hereditary characteristics. Similarly, the resemblance of the isotopic composition of the terrestrial material to that of chondrite suggests that they have a common origin. There are other favorable evidences, and the idea that the earth originated from a substance similar to chondrite is today accepted as one of the promising hypotheses.

On this assumption, the initial quantity of radioactive heat source in the earth can be estimated from the radioactive content of chondritic meteorite (see Table 2-2, page 81). How long ago is it then that the earth came into being?

The Age of the Earth

We have occasionally mentioned the age of the earth as 4.5 billion years. How was this estimated? The disintegration of radioactive substances with long half-lives, such as uranium, thorium, an isotope of potassium (K^{40}), and rubidium (Rb^{87}), serves as the clock. The age of a rock can be estimated by studying the ratio of the parent element and the daughter element of such radioactive substance existing in the rock. In this way, rocks as old as 3 billion years have been found. As the age of the earth must clearly exceed that of rocks preserved on its surface, the earth must be older than 3 billion years.

To determine the earth's exact age, we utilize the isotopic composition of lead. The method is fairly complex so that only a brief outline can be given here. Isotopes of uranium and thorium disintegrate spontaneously and end up as various isotopes of lead as follows:

$$U^{238} \rightarrow Pb^{206}$$
$$U^{235} \rightarrow Pb^{207}$$
$$Th^{232} \rightarrow Pb^{208}$$

Lead, however, has another isotope, Pb^{204}, which is non-radiogenic; that is, it is *not* a product of radioactive disintegration.

At the time of the earth's birth, naturally occurring lead must have contained certain quantities of Pb^{204}, Pb^{206}, Pb^{207} and Pb^{208}. As time went on, the quantity of Pb^{204} must have remained constant while the other three must have increased in quantity because

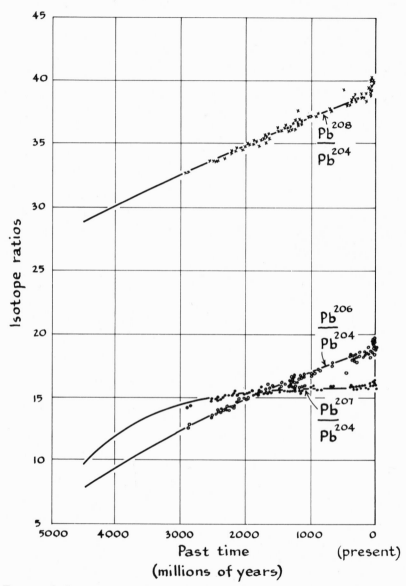

FIGURE 6–6.

Variation of the isotopic composition of lead with age. (From figure 8–8 in *Physics and Geology*, by Jacobs, Russell, and Wilson (1959), with permission of McGraw-Hill Co., New York.)

of the disintegration of uranium and thorium. To determine the isotopic composition of lead at a particular period in the earth's history, all we need to do is to measure the *present* isotopic composition of lead minerals (such as galena), formed at that period. The reasoning is simple; when lead aggregated to form a lead mineral, there is very little chance that either uranium or thorium was included in the mineral; therefore we can assume that a lead mineral formed 500 million years ago, for instance, preserves to this day the "fossil" of the lead isotopic composition prevailing then.

Actual comparison of the isotopic compositions of lead minerals of different ages revealed that Pb^{206}, Pb^{207} and Pb^{208} have indeed increased in quantity (see Figure 6-6). Once the rate of increase is quantitatively determined, the age of the earth can, in principle, be mathematically calculated. In actual practice, the calculation yielded uncertain results because the initial quantities of Pb^{206}, Pb^{207}, and Pb^{208}, existing at the time of the earth's birth, were unknown.

As a solution, C. Patterson, at the California Institute of Technology, adopted the assumption, based on the dust-origin hypothesis, that the earth's original substance was meteorite. Then the quantities of Pb^{206}, Pb^{207} and Pb^{208} existing at the time of the earth's birth can be estimated from the isotopic composition of lead in a meteorite. About 1953, the age of the earth was determined, through this procedure, as 4.5 billion years. This figure coincides with the age of the meteorite itself, determined independently.

Some scientists were dissatisfied with this figure because it is based on the assumption that the terrestrial substance is identifiable with meteorites. By 1956, however, Patterson had greatly improved the accuracy of the experimental technique and proved that the figure of 4.5 billion years can be derived from the terrestrial material alone, without resorting to meteorites. Thus, not only did he confirm the age of the earth as 4.5 billion years, but also corroborated, indirectly, the hypothesis that the terrestrial substance is identifiable with meteorites.

Sand-Sifting Experiment

The factors involved in calculating the age of the earth from the variation of the isotopic composition of lead may be more easily grasped with the aid of an analogy.

Look at Figure 6-7. Suppose there are sand grains of four

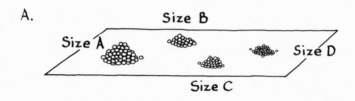

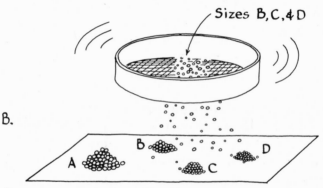

FIGURE 6–7.

Sand-sifting experiment.

different sizes, A, B, C, and D. The whole of Group A and parts of the Groups B, C, and D are placed on a sheet of paper (Figure 6-7A). The remainders of the Groups B, C, and D are thoroughly mixed, placed in a sieve, and sifted onto the sheet of paper on which the whole of Group A and parts of Groups B, C, and D had previously been placed (Figure 6-7B). The sieve can pass grains of all sizes used in the experiment. After an unmeasured period of sifting, the ratios of A, B, C, and D, accumulated on the sheet of paper, are counted. Then follows another period of sifting, succeeded by counting. The procedure is repeated several times.

Now, after the first counting, the experimenter is allowed to have a watch; in other words, from the second round of sifting, he knows the duration of each round, but he does not know the time when he first started sifting. Nor does he know the amount of sand grains from Groups B, C, and D which he placed on the sheet of paper prior to sifting. He does know, however, the rate at which each group of grains, B, C, and D, passes through the sieve; in other words, the only two hints given to the experimenter are the ratio of the four groups accumulated on the sheet of paper at various time-intervals and the rate at which each group falls through the sieve. From these hints, he is asked to calculate the time when he first started sifting.

Essentially that very same problem is involved in the calculation of the earth's age. The three groups of sand in the sieve correspond to U^{238}, U^{235}, and Th^{232}, and the sand grains which accumulate on the sheet of paper correspond to Pb^{206}, Pb^{207}, and Pb^{208}. Group A, the whole of which was placed on the paper, represents Pb^{204}.

The solution of this puzzle is difficult to describe in words, but with the aid of mathematical equations, it is, in principle, quite simple. The mathematical equations are given in the footnote.*

---

* Let the amount of parent and daughter elements at time $t = 0$ be P_0 and D_0 respectively. According to the basic principle of radioactive disintegration, the amount of parent element at time t is

$$P(t) = P_0 e^{-\lambda t}$$

where λ is the disintegration constant. The amount of daughter element at this time, $D(t)$, is the sum of D_0 and the amount produced by disintegration during this period. Thus,

$$D(t) = D_0 + \{P_0 - P(t)\}$$
$$= D_0 + P(t)(e^{\lambda t} - 1)$$
$$\therefore \qquad t = \frac{1}{\lambda} \log e \left\{ 1 + \frac{D(t) - D_0}{P(t)} \right\}$$

In this, λ is a known constant and $D(t)$ and $P(t)$ can be measured. If we know D_0 by some other methods (see page 213), this formula gives the time t.

The Heating of the Earth

Now, if we may assume with some confidence that the earth was formed by an accretion of chondritic substance at a low temperature about 4.5 billion years ago, we ought to be able to estimate its subsequent thermal history. This can be worked out mathematically by assuming that the radioactive heat source was initially distributed uniformly in the earth.

Although the earth was formed in a cold state, some heat must have been generated in the process of accretion, by the compression of the earth's inner part, for instance, or when co mic dust particles collided at high speeds with the half-formed earth. Therefore, how to estimate the earth's initial temperature is one problem. For a mathematical calculation of the earth's thermal history, knowledge of the thermal conductivity of the terrestrial material is essential. Thermal conductivity varies with temperature and pressure. How to estimate this thermal conductivity is the second problem.

There have been many attempts to work out the earth's thermal history but notable results have been obtained only recently with the development of the electronic computer. G. J. F. MacDonald (13) of the United States, E. A. Lubimova (12) of the Soviet Union, and others assumed various sets of values for the initial temperature and the thermal conductivity, and worked out the earth's thermal history for each case.

All the calculations indicate that, even if the earth's initial temperature had been low, the subsequent temperature increase by radioactive disintegration has been considerable (naturally, heat generation was more active in the past, as radioactive elements were more abundant). Figure 6-8 shows the result obtained by Lubimova. The earth's internal temperature, starting at a fairly low value, increased with time till, in approximately a billion years, it reached the melting point of iron at the depth of a few hundred kilometers. Rising still further, it approached the melting point of silicate in two billion years.

If, as the hot-origin hypothesis postulates, the earth was formed by the cooling and the solidification of a ball of hot gas, its initial

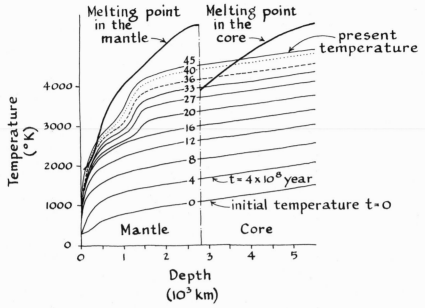

FIGURE 6–8.

Estimated thermal history of the earth. t = time after birth of the earth. Temperature in Kelvin scale ($0°K$ = $-273°C$). (After E. A. Lubimova in *Geophysical J. Roy. Astron. Soc., 1, 1958, 115*, with permission of the author.)

temperature must have been close to the melting point of the initial substance. In that case, the earth, because of subsequent radioactive heating, would have remelted instead of cooling. Once the earth becomes molten, convection would effectively dispose of heat so that the earth would solidify and start heating up again. The earth would be following this cycle of melting and solidification, and could not possibly be cooling progressively as the proponents of the hot-origin hypothesis claimed.

Today, Is the Earth Heating or Cooling?

Thus, according to the dust-origin hypothesis, the initially cold earth has been gradually heating throughout geologic history. Is it still heating today?

An interesting phenomenon in this connection is that the total quantity of the terrestrial heat flow measured at the earth's surface is roughly equal to the quantity of heat that would be generated by a chondrite the size of the earth.

What does that imply? If the hypothesis of a chondritic earth is correct, the earth is losing as much heat as it is generating; hence there is no heat available for warming the earth. In other words, we can say that today, the earth is neither heating nor cooling. This is the answer—a somewhat unexpected one—to the questi n we posed at the beginning of the chapter. The initially cold ea h had been gradually heating up, but at the present moment, it se ms to have reached a state of thermal balance.

As the stock of radioactive elements decreases with time, the earth will no doubt eventually start cooling.

Formation of the Core

One point on which the dust-origin hypothesis was formerly considered to be weaker than the hot-origin hypothesis was in the explanation of the layering of the earth into a core, a mantle, and a crust. In the case of the hot-origin hypothesis, it is highly conceivable that while the earth was still a hot fluid, the heavy iron settled to the center, over which the lighter silicate formed a layer (the mantle), and the lightest substance, the crustal material, rose to the surface. If, however, meteoritic substances aggregated in a cold state to form the earth, how can we explain the separation of iron and silicate?

Fortunately, according to the earth's thermal history explained above, the initially cold earth gradually heated up till, in about a billion years, its temperature reached the melting point of iron at the depth of a few hundred kilometers. At that time, iron may have melted and sunk to the center to form the core. Later, in about two billion years, the temperature in the upper mantle (at the depth of about 1000 km) approached the melting point of silicate. Then, it is possible that the light crustal material, rich in radioactivity, rose to the surface. This would explain why radioactive substances are today mostly concentrated in the crust.

The main difference from the hot-origin hypothesis is that, according to the cold-origin hypothesis, the layering process has been proceeding slowly throughout geologic history and may be going on even today. This idea of the continually "developing" earth is in strong contrast to the traditional belief that the earth, having started as a ball of hot gas, is gradually advancing towards the cold of "death."

Growth of the Core and Continental Drift

The idea of the growth of the core has a bearing on our main theme, continental drift. The leading English paleomagnetist, Runcorn, connected this idea with the mantle convection theory and in 1962 proposed a new hypothesis in support of continental drift (16).

Two serious objections against the theory of continental drift were (1) the lack of an adequate mechanism for the drift and (2) the unnaturalness of the recent occurrence (Mesozoic) of the drift. The first objection can be dismissed if there is convection in the mantle. In answer to the second objection, Runcorn proposed the following idea.

According to the dust-origin hypothesis, the core has been growing slowly throughout the earth's history. Hence, in remote antiquity, the earth had a tiny core, and its size has been growing with time. What kind of mantle convection would occur in such an earth? While the core was tiny, a single large current involving the whole mantle is likely to have occurred (see Figure 6-9A). During such period, the light crustal material which had risen to the surface could have been gathered by the current into one huge continent. As the core grew in size (see Figure 6-9B), the mantle would become less thick, and hence no longer able to maintain a single large current; accordingly, several separate currents would be set up in its place. This would tend to pull apart the single continent, and the separated land masses would be carried by the new currents to the points where the currents turn down. Upon further growth of the core, the currents would become even smaller and

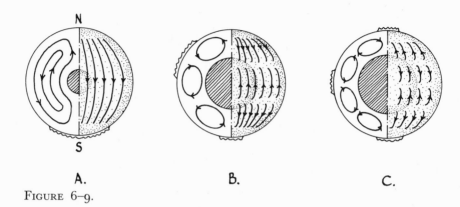

FIGURE 6–9.

Convection in the mantle with different sizes of the core. (From figure 26 in *Continental Drift*, edited by S. K. Runcorn (1962), with permission of Academic Press Inc., New York.)

hence there would be further splitting of land masses (Figure 6-9C). If continental drift was caused by this change in the pattern of convective currents, it is natural that the drift should have occurred fairly recently, that is, with the growth of the core.

Runcorn's idea was based on the work of S. Chandrasekhar, a great Indian astrophysicist in the United States, who had worked out mathematically the relationship between the size of convective currents and the size of the core. According to Chandrasekhar, the change in the size of the currents due to the growth of the core is a discontinuous process; when the core reaches a certain size, the currents which have been flowing steadily change abruptly into smaller currents. Runcorn showed that if we assume a certain reasonable value for the growth rate of the core, "the continental drift in the Mesozoic" can be explained as the result of the change from (B) to (C) in Figure 6-9. Runcorn's idea, which cleverly combines the new theory on the earth's origin with the theory of mantle convection in explanation of continental drift, may sound too ingenious, but nevertheless it offers an interesting possibility.

7

Hint from the Ocean Floor

Study of the Ocean Floor

More than two-thirds of the earth's surface is covered by the sea. The study of the ocean is hence indispensable for a correct understanding of the earth, and oceanography occupies an important place in earth science. Traditionally, the sea water had received greater attention in oceanography than the ocean "floor," partly because geoscientists, busy with research on land, had not had time to spare for the ocean floor, and partly because there had not been adequate means for exploring the mysteries of what lies beneath the oceans.

After World War II, however, the spectacular development of observational techniques opened a new vista in this field. The traditional picture of the ocean floor as a featureless flat basin was overthrown by the discovery of submarine volcanoes whose height surpasses that of Mt. Fuji (3776 m), and of rugged submarine ranges, stretching for thousands of kilometers. Geoscientists came to realize that they could not discuss the earth further without adequate knowledge of the sea floor.

So far, extensive exploration of the ocean floor has been carried out mainly by the United States, U.S.S.R., Great Britain and Japan. In these countries, most of the observational techniques that had been used ashore were successfully adapted for use at sea. Typical ones are the exploration of the submarine crustal structure through artificial earthquakes (deep-ocean seismic sounding); the development of ship-borne magnetometers and gravity meters; and the measurement of heat flow through the ocean floor (see page 193). Another important technique is the precise echo-sounding method for determining the topography of the ocean bottom; periodic signals are sent to the ocean bottom and the times required for their echoes to return are automatically recorded and interpreted as depth profiles.

Once the technique is established and the vessel constructed, research is, in many respects, easier at sea than on land. On land, we often encounter practical difficulties in reaching the spot we wish to survey: the red tape involved in crossing a nation's border, the transportation of heavy instruments, and so on. At sea, once the instruments are taken aboard, the whole ocean is at our disposal.

An important result of exploration through such techniques was the establishment of the nature of the suboceanic crustal structure which was explained in Chapter 1 (see Figure 1-13, page 39); the suboceanic crust was found to be only a few kilometers thick and thus differs markedly from the continental crust. Another remarkable achievement was the measurement of heat flow through the ocean floor, mentioned in the previous chapter.

It is beyond the scope of this book to enumerate all the recent findings of the physics of the ocean floor. We will only take up a few topics relevant to continental drift.

The Crust Has Actually Been Displaced

The theory of continental drift postulates a displacement of the crust for thousands of kilometers. Although there are many indirect evidences of such displacement, the only direct evidence of extensive horizontal movement of the crust has been the existence of great faults at various places in the world. The most spectacular of these is the San Andreas Fault along the west coast of North

America. (See Figure 7-1.) This fault runs along western California, from the northwest to the southeast, for more than 960 km. According to geological evidences, the east side of this fault has, since the Tertiary, been offset 200 km in the southeastern direction relative to the west side, and the displacement still goes on today, at the rate of about 5 cm per year.

FIGURE 7-1.

Map showing schematically the position of San Andreas Fault and the East Pacific Rise. The arrows indicate the directions of the fault movement.

Recently, great faults surpassing the San Andreas Fault in scale, have been found at sea by R. Mason, A. Raff, V. Vacquier (*1*, *16*), and others at the Scripps Institution of Oceanography, University of California. These scientists carried out a detailed magnetic survey of the East Pacific, and found that the geomagnetic anomaly in this area shows quite an orderly pattern; the pattern consists of lineations, several tens of kilometers apart, trending in the north-south direction. Such an orderly, lineated pattern has never been observed on land and is in itself a significant discovery: it evidently indicates some unusual structure in the ocean floor. What is more, the magnetic survey shed an interesting light on possible crustal displacement. Figure 7-2 shows a part of the results obtained by these scientists. We can see that there is a distinct displacement of the pattern along a line which runs east-west at the latitude of 34°N. The pattern can be matched by shifting the southern portion 135 km eastward. This obviously

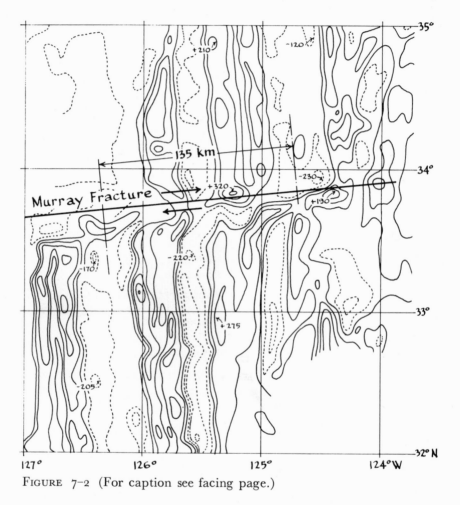

FIGURE 7-2 (For caption see facing page.)

indicates that, *after* the formation of the geomagnetic anomaly pattern, a fault movement along this line caused an east-westerly displacement of 135 km. This interpretation is borne out by the fact that, topographically, a fault is known to exist at this locality (the Murray Fracture off the west coast of California). Encouraged by this discovery, these scientists made a similar magnetic survey along other known faults of the East Pacific. The result was that east-westerly displacements were found along all faults, the largest one being the 1150-kilometer offset along Mendocino Fracture which strikes east-west at the latitude of 40°N.

The evidence of large-scale horizontal displacement along a fault is of great significance for the continental drift theory. If the suboceanic crust moves that much, why not the continental crust?

Rifts of the Earth

A great submarine range, rising over 3000 m above the ocean floor, almost bisects the Atlantic from north to south. This is the so-called Mid-Atlantic Ridge. M. Ewing, B. Heezen, and others (5, 16) at Lamont Geological Observatory, Columbia University, explored the Ridge in detail and made an interesting discovery. The Ridge, approximately 1000 km wide, rises gently from the ocean floor (5000 m deep), but the central zone (about 200 km wide) is marked by rugged mountainous topography with peaks well over 1000 m high. Incised into this central zone, they found a narrow rift valley, about 15 km wide, running along the crest of the Ridge in the north-south direction. The profile of the Ridge obtained by these scientists is shown in Figure 7-3. They interpreted this central rift valley as a rift of the crust.

FIGURE 7–2.

Map of the anomaly of the total intensity of the geomagnetic field in the East Pacific Rise. The unit is 10^{-5} gauss. The arrows indicate the directions of fault movement. The solid contour indicates positive anomaly, and the dotted contour negative anomaly. (After figure 2 in A magnetic survey off the west coast of the United States between lat. 32° & 26° N, long. 121° & 128° W., by R. Mason, in *Geophys. J. 1.*, *p. 320*, *1958*, with permission of the author.)

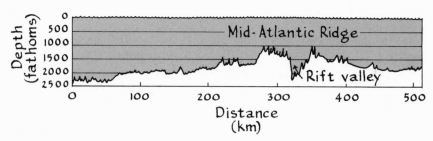

FIGURE 7-3.

Profile of the Mid-Atlantic Ridge. (From figure 20 in *Continental Drift*, edited by S. K. Runcorn (1962), with permission of Academic Press Inc., New York.)

What is more, they found that submarine mountain ranges like the Mid-Atlantic Ridge exist in every ocean of the world, and in fact form a world-girdling system as shown in Figure 7-4. The Mid-Atlantic Ridge, extending southward, circles around to the south of the African continent and enters the Indian Ocean. In

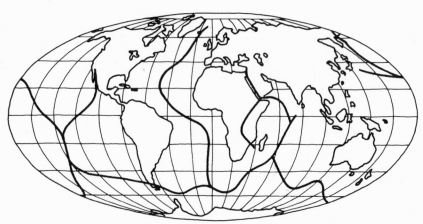

FIGURE 7-4.

World-girdling ridge system. (From figure 28 in *Continental Drift*, edited by S. K. Runcorn (1962), with permission of Academic Press Inc., New York.)

the middle of the Indian Ocean, the Ridge bifurcates and its eastern branch extends through the Antarctic Sea into the Pacific where it joins the well-known East Pacific Rise, off the west coast of South America. Not all the ridges are marked by a central rift valley; the East Pacific Rise, for instance, is a low smooth bulge, and has no central rift along its crest. Nevertheless, it has been suggested that these world-wide mid-oceanic ridges have some common origin, and evidence is accumulating that these ridges represent real rifts of the earth's crust.

For one thing, numerous earthquakes occur along the crest of these ridges. Generally, earthquakes begin in a very small area; the point at which the first movement seems to occur is called the *focus*. The point on the earth's surface directly above the focus is called the *epicenter*. Epicenters of earthquakes occurring near the Mid-Atlantic Ridge are plotted in Figure 7-5. It is clear at a glance that they are precisely aligned along the crest of the Ridge. A similar correspondence was found in the East Pacific Rise. Evidently, there is some activity going on beneath these ridges.

Heat flow measurement provides another evidence that these ridges represent cracks in the earth's crust. Extensive measurement in the East Pacific, conducted by the Scripps Institution of Oceanography, revealed that the heat flow is abnormally high along the crest of submarine ridges. The result of the survey, shown schematically in Figure 7-7, is striking. At the crest, the heat flow is 8 times the normal value, while along the flanks of the ridge, the flow is abnormally low. Similar results have been obtained for the Mid-Atlantic Ridge. We saw in Chapter 6 that the equivalence of the average heat flow at land and sea was difficult to explain in the absence of adequate heat supply in the suboceanic crust (page 195). As a solution, it was suggested that heat is generated in the suboceanic mantle and transferred to the surface by mantle convection. This time, heat flow many times the normal value has been observed at sea along the crest of ridges. Naturally it was inferred that the abnormally high temperature that prevails beneath the crest is maintained by the mechanism of mantle convection.

There is another line of evidence for the abnormally high tem-

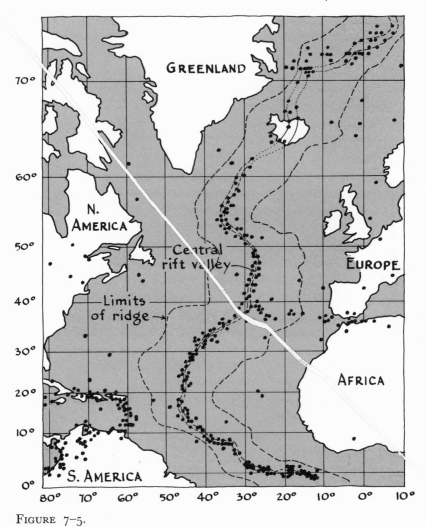

FIGURE 7-5.

Epicenters of earthquakes along the Mid-Atlantic Ridge. (From figure 17 in *Continental Drift*, edited by S. K. Runcorn (1962), with permission of Academic Press Inc., New York.)

perature under the ridges; according to deep-ocean seismic sounding, seismic waves travel at an abnormally low speed in the upper mantle beneath the East Pacific Rise. Other things being equal, the speed of seismic wave decreases with increasing temperature.

FIGURE 7–6.

Heat flow measurement in Southeastern Pacific.

Therefore an abnormally low speed implies an abnormally high temperature.

These phenomena can be interpreted as follows. Look at Figure 7-8. It is possible that the bulging mountainous topography of these mid-oceanic ridges and rises was produced by the upward limb of a rising mantle convection current. Naturally, because of the high temperature of the mantle substance, abnormally high heat flow prevails in the central zone of the ridge. The zone is also marked by occasional volcanic activities which produce rugged volcanic topography. Observed extremely high heat flow on the crust is sometimes too localized to be caused directly by the mantle current. Such high heat flow is interpreted as due to volcanic activities. When the upwelling current runs up against the crust, it divides into two, pulling apart the crust and producing a central rift valley such as that seen in the Mid-Atlantic Ridge. If this interpretation is correct, the mid-oceanic ridge provides another evidence in favor of the mantle convection theory.

The reason for the absence of the central rift valley in the East Pacific Rise is not known; some scientists attribute it to the relative youthfulness of the formation. The important and radical point about these ridges is that while continental mountain ranges are

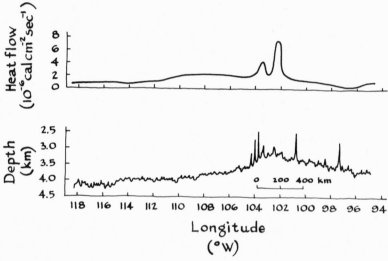

Figure 7–7.

Profile of heat flow and topography across the East Pacific Rise. (After figure 4 in Heat flow through the Eastern Pacific Ocean Floor, by R. P. Von Herzen and S. Uyeda, in *JGR. 68, 4219, 1963*, with permission of the authors.)

considered to be the product of folding due to compressional forces, the origin of submarine ridges is attributed to tensile forces which produce rifts.

If convective currents do indeed produce rifts in the earth, it is tempting for the advocates of the continental drift theory to consider these currents as the mechanism of the continental split and the subsequent drifting.

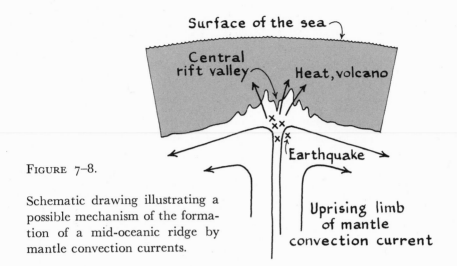

FIGURE 7–8.

Schematic drawing illustrating a possible mechanism of the formation of a mid-oceanic ridge by mantle convection currents.

Are Continents Being Rifted Apart?

Turning back to Figure 7-4, we can find two regions where the world-encircling ridge-rift system impinges upon continents. One is the Mid-Indian Ridge which enters via the Gulf of Aden into the Red Sea (see Figure 7-9). The southern branch of this Ridge appears to join the rift valleys of East Africa. This region, characterized by a chain of deep lakes trending north-south is known as the East African Rifts. The other is the East Pacific Rise which trends northward from Easter Island, approaches the North American Continent and enters the Gulf of California. An English scientist, R. Girdler, suggested in 1962 that in these regions, the continental mass is actually being rifted apart by mantle convection currents (1).

A number of geophysical data support this idea. In the central region of the Red Sea, there are large anomalies in the intensity of the geomagnetic field, and the trend of anomalies is almost exactly parallel to the shorelines (see Figure 7-10). Such anomalies are often caused by local excesses of basaltic rocks which are highly magnetic. Girdler suggests that, in this region, the continental crust has been torn apart, leaving room for the intrusion of basaltic rocks. This interpretation is borne out by seismic evidence. Near the center of the Red Sea, the speed of the P wave is about 7 km/sec at the relatively shallow depth of about 4 km. This speed is characteristic of basaltic rocks (see page 37). Evidently, in this region,

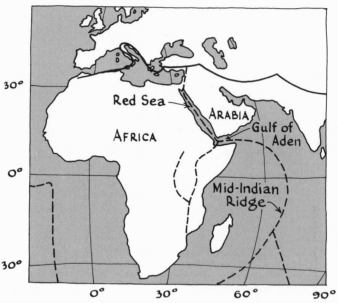

FIGURE 7–9.

Map showing schematically the position of the Mid-Indian Ridge.

the basaltic layer comes up fairly close to the surface of the sea floor. The Gulf of Aden and the Red Sea, moreover, are characterized by an abnormally high heat flow.

The same can be said of the Gulf of California. Geological evidence indicates that Baja California (Figure 7-1) once formed a part of the mainland and has since rifted away. Seismic sounding testifies that below the Gulf of California the Mohorovičič discontinuity lies only about 6 km beneath the sea floor. Therefore, the Gulf, like the Red Sea, has a crustal structure which is essentially oceanic in character; in other words, these narrow belts of

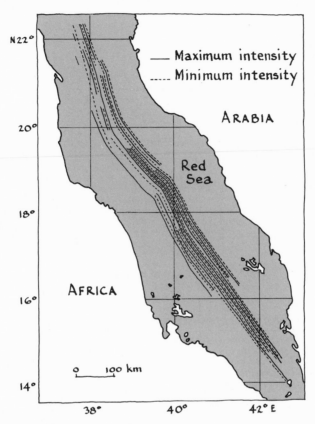

FIGURE 7-10.

Anomaly of the total intensity of the geomagnetic field in the Red Sea. Lines of maximum and minimum total intensity are shown. (Adapted from Initiation of continental drift, by R. W. Girdler, in *Nature, Vol. 194, No. 4828, 1962, p. 521–524,* with permission of the author.)

water represent true seas in the process of formation, and not parts of continents that have been submerged. Heat flow in the southern part of the Gulf of California is many times higher than the normal value.

All these evidences indicate the possibility that the continental mass is actually being torn apart in these regions.

Is the Atlantic a Rift?

Recently, J. T. Wilson at the University of Toronto, Canada, has advanced a hypothesis in support of continental drift (*22*). Since Wilson's hypothesis is an eloquent and overall exposition of the idea of the continental drift by mantle convection, which has been enjoying more and more support recently because of accumulating evidence, we will introduce his idea by way of rounding off this long history of the continental drift theory.

If the Red Sea and the Gulf of California represent newly developing rifts in the continental mass, the Atlantic Ocean can be regarded as a gigantic rift fully developed. Rising mantle convection currents, separating beneath the primitive continent *Pangea*, pulled it apart to form the Euro-African Continent and the two Americas. As Holmes suggested, the convective currents constitute a kind of belt conveyer on which the split land masses are carried away from each other. The widening rift is filled by the mantle substance (brought up by ascending currents) which pours out to form the mid-oceanic ridge. The high temperature of such substance upwelling from the depths of the mantle accounts for the high heat flow along the crest of mid-oceanic ridges. The mantle substance, in rising to the surface, escapes the high pressure prevailing in the earth's interior; hence, its melting point decreases and a part of the mantle substance melts into lava. Such lava produces volcanoes as well as the submarine crust. The process is schematically illustrated in Figure 7-11A. The newly formed crust and the volcanoes overlying it are gradually carried away east and west by the bifurcated currents. (See Figure 7-11B.) According to this interpretation, the volcanic islands now scattered over the Atlantic were originally formed at the mid-oceanic ridge, and are now "travelling" in the wake of continents.

When an advancing continent meets a descending current, the forward movement is necessarily stopped and the light continental crust piles up at the front margin to form mountains, while the ocean floor is pulled down by the descending currents to form a trench. The Andes range is considered to have originated from such an encounter between the American Continent (advancing

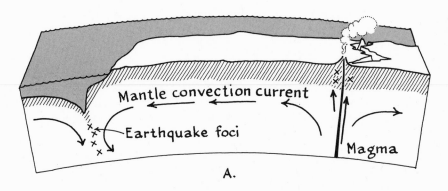

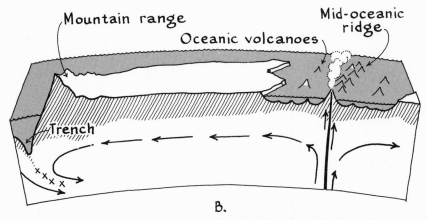

FIGURE 7–11.

Schematic drawing illustrating Wilson's interpretation of the formation of mid-oceanic ridges and continental drift by mantle convection. (After figure on p. 8 in Continental drift, by J. T. Wilson, *Scientific American*, *April 1963*, with permission of the author.)

westward) and the eastward current under the Pacific Ocean between the East Pacific Rise and the west coast of the continent

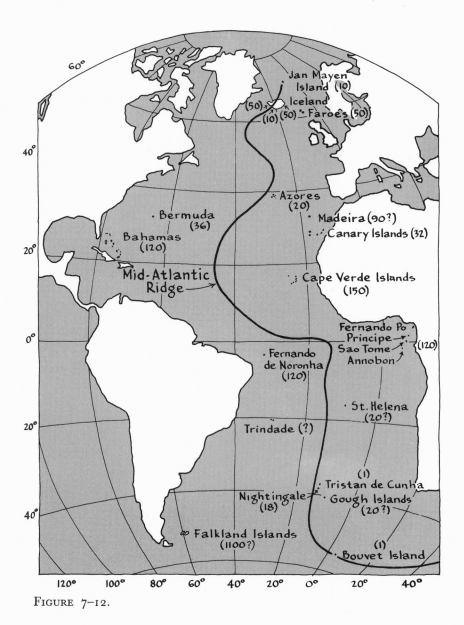

FIGURE 7-12.

Age of Atlantic islands, as indicated by the age of the oldest rocks found in them. The numbers associated with the islands give ages in millions of years. (After figure on p. 4 in *Continental drift*, by J. T. Wilson, in *Scientific American, April 1963*, with permission of the author.)

Speed of the Belt Conveyer

If, as Wilson claims, the volcanic islands of the Atlantic were originally formed at the mid-oceanic ridge and are now being carried away on the common conveyer of convection currents, then the age of the islands should increase with increasing distance from the ridge. The age of the islands can be determined by the radioactive dating of igneous rocks found there. Incidentally, the age determination of an isolated island, lacking sedimentary layers and hence fossils of ancient organisms, was quite impossible before the establishment of the radioactive method. The development of a new measurement technique not only helps to increase data but also opens up new vistas in science.

The ages of the islands, thus determined, are shown in Figure 7-12. While the islands close to the Mid-Atlantic Ridge are relatively young (10 million years for Iceland, 20 million years for the Azores), the islands grow progressively older with the distance from the Ridge (36 million for Bermuda, 50 million for the Faroe Islands, and 120 million for Fernando-Po and Principe situated off the west coast of Africa). The result is in excellent agreement with the postulate of the continental drift theory that the Atlantic began to open up in the Jurassic (approximately 200 million years ago).

The conveyer's speed, calculated from the ages of the islands and their distances from the ridge, is $2 \sim 6$ cm per year. This is in keeping with the estimated speed of continental drift and of the mantle convection.

Wilson's idea gives us a new image of the ocean bottom; the ocean floor is not an old and unchanging formation but, on the contrary, ever being formed anew, moving and ultimately descending into the mantle. This interpretation sheds light on two mysteries that had baffled oceanographers. One is that no rock or sediment older than the late Mesozoic has ever been dredged from the ocean floor. The other is that, according to deep-ocean seismic sounding, the layer of sediments covering the ocean floor is only several hundred meters thick; if the ocean floor had existed from remote antiquity, the thickness ought to have been about ten times greater. Both mysteries are solved if the sea floor is an ever-changing, young formation.*

The Current in the Pacific

So much for the Atlantic. What about the Pacific Ocean? Turn to Figure 7-13. A careful observer would notice that the chains of volcanic islands in the Pacific all trend in one direction, from the southeast to the northwest. The Hawaiian Islands are a good example. Geologic studies revealed that the islands of this chain are oldest at the northwest end and grow steadily younger toward the southeast. The Island of Hawaii at the southeast end has an active volcano that is still erupting. Wilson's explanation for this phenomenon is easily grasped from Figure 7-14. Lava flows, ascending from the depths of the mantle, erupt to form active volcanoes. The convective current in the upper mantle, which is relatively rapid, conveys the crust northwestward. With the crust, the volcanic pile is carried away from the source of lava flow towards the distant sea, while new volcanoes form over the source.

This interpretation suggests that the mantle convection current under the Pacific flows from the southeast to the northwest direction. Presumably, this is the current which ascends at the East Pacific Rise, flows across the Pacific (in the direction of the arrows in Figure 7-13) and descends beneath the Asian Continent. The descending current may have something to do with the origin of island-arcs and trenches (including the Japan trench) found along the eastern fringe of the Asian Continent. Thus, the topic turns upon the Japanese Islands.

---

* However, there are also some evidences against the youthfulness of the ocean floor. Recently, M. Ewing and his colleagues have undertaken a thorough investigation of the sediments of the ocean bottom by seismic profiler. They are finding many important pieces of evidence, which, according to them, testify to the *oldness* of the ocean floor.

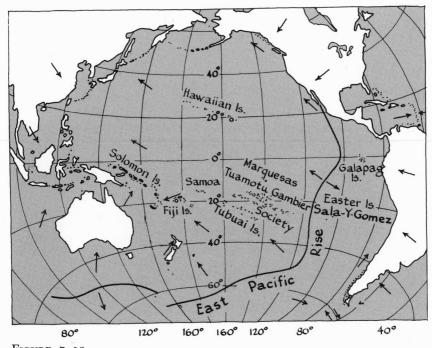

FIGURE 7–13.

Pacific islands. The arrows show the probable directions of convection current flow. (After figure on p. 14 in *Continental drift*, by J. T. Wilson, in *Scientific American, April 1963*, with permission of the author.)

Below the Japanese Islands

Japan, known for her earthquakes and volcanic activities, is an exciting place to explore geologically. This island-arc presents a rich variety of geologic activities, and the geology of the country would itself warrant a book. Here, on account of limited space, we will confine ourselves to a few topics relevant to our subject.

If, as the mantle convection theory suggests, the Japan trench marks the place where the currents descend into the mantle, pulling down the ocean floor, then, in the trench, the heat flow ought to be below the normal value. In Japan, heat flow has been measured in considerable detail over land and sea. Figure 7-15 shows the summary of results obtained so far. The distribution is remarkable; heat flow is abnormally low in the crosshatched areas, including the Japan trench, the Kurile trench and a portion of the Pacific off the northeastern coast of Japan.

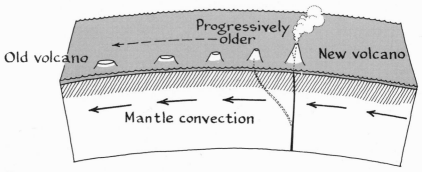

FIGURE 7–14.

Schematic drawing illustrating a possible mechanism of the formation of volcanic-island chains. (After figure on p. 9 in Continental drift, by J. T. Wilson, in *Scientific American, April 1963*, with permission of the author.)

Related to this fact is the distribution of earthquake foci in Japan. There are many kinds of earthquakes, and they cannot be discussed in a lump. However, it is a well-established fact that the foci of earthquakes are shallow on the Pacific side of Japan (including the Japan trench) and grow deeper toward the Asian Continent. Schematically, the foci seem to be distributed along a plane dipping towards the Asian Continent (see Figure 7-16). If we assume that this imaginary plane represents a sliding surface where the convection current, meeting the continental mass, plunges downward, then we can interpret the occurrence of earthquakes along this plane as due to the great force at work there. Earthquakes along this plane sometimes reach the focal depth of several hundred kilometers and are hence called "deep-focus earthquakes." Strangely enough, deep-focus earthquakes are restricted to regions around the Pacific, forming the so-called circum-Pacific belt; there,

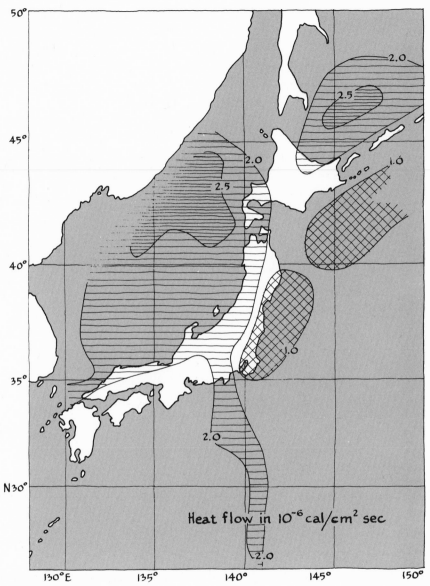

Distribution of heat flow in and around Japan. Crosshatched areas indicate regions of low heat flow and horizontally hatched areas those of high heat flow. (After V. Vacquier, S. Uyeda, M. Yasui, and T. Watanabe, 1966, with permission of the authors.)

they are known to occur invariably along planes dipping toward the continental masses.

Beside the zones of low heat flow mentioned above, there are, in Japan, zones of abnormally high heat flow, marked with horizontal hatches in Figure 7-15. Significantly, almost all the volcanoes of post-Tertiary origin are found in the latter zone. This

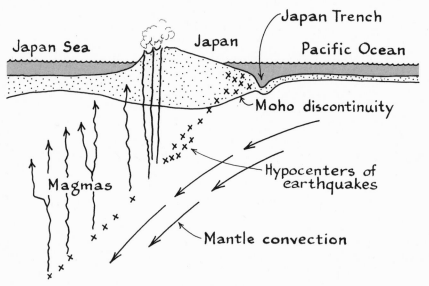

FIGURE 7–16.

Sketch showing schematically the distribution of earthquake foci beneath Japanese Islands.

indicates that in Japan high heat flow is closely associated with volcanic activities. According to the mantle convection theory adopted here, the convective currents plunge *downward* beneath this volcanic region. Earlier in this chapter, we discussed the formation of volcanoes over *rising* mantle convection currents; the rising mantle substance melts (because of decreasing pressure) into magma and pours out to form volcanoes. In Japan, by contrast, volcanic activities seem to occur along descending currents. If both are correct, the magma, the source of volcanic activity, is supplied in two ways, either by the ascending currents through the ocean floor or by descending currents at the continental margin. It has been suspected, from the petrologic point of view, that the basaltic magma from the oceanic volcano and the basaltic or andesitic magma from the volcano at the continental margin are formed in different ways. If so, the difference in origin, explained above, might provide a clue. The theories of mantle convection and continental drift have bearing on such questions as well.

Heat flow is high in both the Japan Sea and the Sea of Okhotsk. Since both these marginal seas have a typically oceanic crust, i.e., thin basaltic crust, high heat flow must indicate some unusual situation in the mantle. We suspect that igneous activities, which do not come up to the surface, may be going on there. Some scientists are inclined to consider that the Japan Sea is also a rift, just like the Gulf of California, and that the Japanese Islands have drifted away from the Asian Continent.

A Remark on the Mantle Convection Theory

It must be remembered that Wilson's explanation of continental drift and the above interpretation of seismic and heat flow data in Japan are entirely based on the hypothesis that convection actually occurs in the mantle. We have seen, throughout the book, that many of the phenomena observed at the earth's surface can be satisfactorily explained *if* convection occurs in the mantle: the periodic occurrence of orogeneses (page 87); the equivalence of heat flow through continental rocks and ocean basin rocks (see page 195); and the origin of mid-oceanic ridges and rifts.

The isostatic balance that prevails between the crust and the mantle suggests that the mantle substance can behave as a fluid. If the deeper part of the mantle contains even a small amount of radioactive substance, it is possible that a slow convective current starts in the mantle. However, there has as yet been no *direct* proof that such current *does* actually exist in the mantle. The hypothesis of mantle convection, like so many other hypotheses in earth science, requires further proofs before it can be definitely established.

8

Continental Drift in the Modern Light

Revolution in Earth Science

The history of the continental drift theory is marked by vicissitudes. The similarity of the Atlantic coastlines first suggested the idea, and geological, paleontological, and paleoclimatological studies revealed evidences in favor of it. Objections were raised and the first round of controversy ended inconclusively. After World War II, the controversy was rekindled through the study of rock magnetism. This time, more concrete evidences were offered, and in addition, a firmer theoretical background provided by the mantle convection theory. Today, the fast developing science of the ocean floor is revealing fresh evidences in favor of the idea of drift.

It is Wilson's belief:

. . . that earth science is ripe for a major scientific revolution, that in a lesser way, its present situation is like that of astronomers before the ideas of Copernicus and Galileo were accepted, like that of chemistry before the ideas of atoms and molecules were introduced, like that of biology before evolution, like that of physics before quantum mechanics. Before each revolution, all the pegs seemed square and all the holes round. In each case, it was not

until it was realized that one had to discard the whole frame of
reference and seek another that answers came in a flood. If earth
scientists have been trying to fit the history of the earth which
has in truth been mobile into the framework of a rigid and fixed
pattern of continents, then it is not surprising that it has been
impossible to answer the major questions. It is not our methods
nor our observations that have been wrong, but our whole
attitude.

All of these major revolutions cast their shadows a long way
before them. Greek astronomers foreshadowed Copernicus by
2,000 years; Darwin got ideas about evolution from his grand-
father. It should not surprise us that Wegener's and Holmes'
ideas have been before us for some time without acceptance. (*23*)

Today, the consensus of opinion seems to be moving towards
reconsideration, if not acceptance, of Wegener's and Holmes' ideas.
One recent evidence of this is the new Symposium on Continental
Drift, held in London in 1964. The symposium brought together
many leading scientists from all over the world working in many
branches of geophysics. Some of the participants whose names are
already familiar to us were Blackett, Runcorn, Bullard, Wilson,
Vacquier, Heezen, and Jeffreys. The main topics at the symposium
concerned new evidences from paleomagnetism and oceanography
(such as those we discussed in a simplified form in the latter half
of this book) and the controversial problem of the physics of convection currents in the mantle.

In the Introduction to the Symposium, Blackett remarked that
the evidence of great crustal movement such as is seen in the San
Andreas Fault suggests that, today, the question is no longer the
qualitative one, "Have or have not the continents drifted?" but
the quantitative one, "How much have they drifted and when?".
Blackett adds:

The new evidence acquired during the last decade from oceanography and rock magnetism will, I think, appear in the history of
the subject as supporting and making more quantitative an
already rather strong qualitative case. (*1*)

Concluding Remarks

Continental drift is the leitmotif of this overture to the study of earth science. We have frequently digressed and played variations on our theme, yet scientific exploration itself usually follows just such an intricate winding pattern. The work of Wegener, Holmes, and Blackett not only teaches us about continental drift but also gives us a glimpse of the problems, joys, and disappointments that inhere in scientific progress.

We have intentionally covered a wide range of subjects relevant to continental drift and the physics of the earth. Yet the earth retains many secrets. As Isaac Newton once remarked:

> I seem to have been only like a boy playing on the seashore, and diverting myself in now and then finding a smoother pebble or a prettier shell than ordinary, while the great ocean of truth lay all undiscovered before me. (7)

Will the theory of continental drift and the debate about it prove to be merely a pebble? Or is it a shell significant of the fact that we are on the shore of an ocean of truth? No one yet knows. Nor is it as important that we answer that question, as it is important that we continue to debate and to search for truth.

References

1. *A Symposium on Continental Drift.* London: The Royal Society, 1965.

2. Bullard, E. C. and Gellman, H. "Homogeneous dynamos and terrestrial magnetism," *Phil. Trans. Roy. Soc.*, London, Ser. A, 247, 1954, pp. 213–78.

3. Doell, R. R., Cox, A. and Dalrymple, G. B. "Reversals of the earth's magnetic field," *Science*, 144, 1964, pp. 1537–43.

4. Du Toit, A. L. *Our Wandering Continents.* Edinburgh: Oliver and Boyd, 1937.

5. Hill, M. N. (ed.). *The Sea—ideas and observations on progress in the study of the seas.* Vol. 3. New York and London: Interscience Publisher, 1963.

6. Holmes, A. *Principles of Physical Geology.* London and Edinburgh: Nelson, 1945.

7. Hutchins, R. M. (ed.). "Biographical Note to the Mathematical Principles of Natural Philosophy," *Great Books of the Western World*, 34, Newton, Huygens. Encyclopaedia Britannica, Inc., 1952.

8. Irving, E. *Paleomagnetism.* New York, London and Sydney: John Wiley & Sons, 1964.

9. Jacobs, J. A., Russell, R. D. and Wilson, J. T. *Physics and Geology.* New York, Toronto and London: McGraw-Hill, 1959.

10. Jeffreys, H. *The Earth.* 4th ed. Cambridge: Cambridge University Press, 1959.

11. Joly, J. *Surface History of the Earth.* Oxford: Oxford University Press, 1930.

12. Lubimova, E. A. "Thermal history of the earth with consideration of the variable thermal conductivity of the mantle," *Geophy. J.* 1. 1958, pp. 115–34.

13. MacDonald, G. J. F. "Calculation on the thermal history of the earth," *J. of Geophysical Research*, 64, 1959, pp. 1967–2000.

14. Nagata, T. *Rock Magnetism.* Revised ed. Tokyo: Maruzen, 1961.

15. Rikitake, T. *Electromagnetism and the Earth's Interior.* Amsterdam, London and New York: Elsevier, 1966.

16. Runcorn, S. K. (ed.). *Continental Drift.* New York and London: Academic Press, 1962.

17. ———. "Rock Magnetism—Geophysical Aspects," *Phil. Mag., Supplement*, 4, 1955.

18. Scheidegger, A. E. *Principles of Geodynamics*. 2nd ed. Berlin, Göttingen and Heidelberg: Springer, 1962.

19. *Theory of Continental Drift, A Symposium*. Tulsa: American Association of Petroleum Geologists, 1928.

20. Verhoogen, J. "Temperature within the earth," *Physics and Chemistry of the Earth*. London, New York and Paris: Pergamon, 1956, pp. 11–43.

21. Wegener, A. *The Origin of Continents and Oceans*. London: Methuen, 1924.

22. Wilson, J. T. "Continental Drift," *Scientific American*, April, 1963.

23. ———. "The Movement of Continents," *Symposium on the Upper Mantle Project*, International Union of Geodesy and Geophysics, XIII General Assembly, Berkeley, 1963.

Index